Life in Space

Life in Space

by Dinah L. Moché

A Ridge Press Book
A & W Visual Library, New York

Editor-in-Chief: Jerry Mason
Editor: Adolph Suehsdorf
Art Director: Harry Brocke
Managing Editor: Marta Hallett
Associate Editor: Ronne Peltzman
Art Associate: Nancy Mack
Art Associate: Liney Li
Art Production: Doris Mullane
Picture Editor: Marion Geisinger
Art Associate: Penny Burnham
Production Consultant: Arthur Gubernick

Title page: Unmanned Viking
in orbit around Mars.
Contents page: Artist's concept
of Saturn from satellite Rhea.

First published in the United States of America
in 1979 by A & W Publishers, Inc.
95 Madison Avenue
New York, New York 10016
By arrangement with The Ridge Press, Inc.

Library of Congress Catalog Card Number: 79-51841
ISBN: 0-89104-155-9   (paperback)
ISBN: 0-89104-154-0   (hardcover)
Printed in the United States of America

To my parents
Mollie and Bertram Levine,
who showed me the joy
of lifelong learning,
with love and appreciation.

Contents

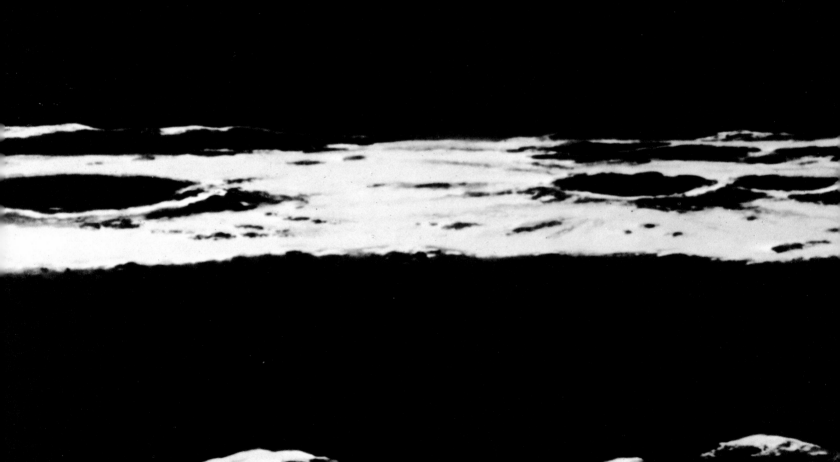

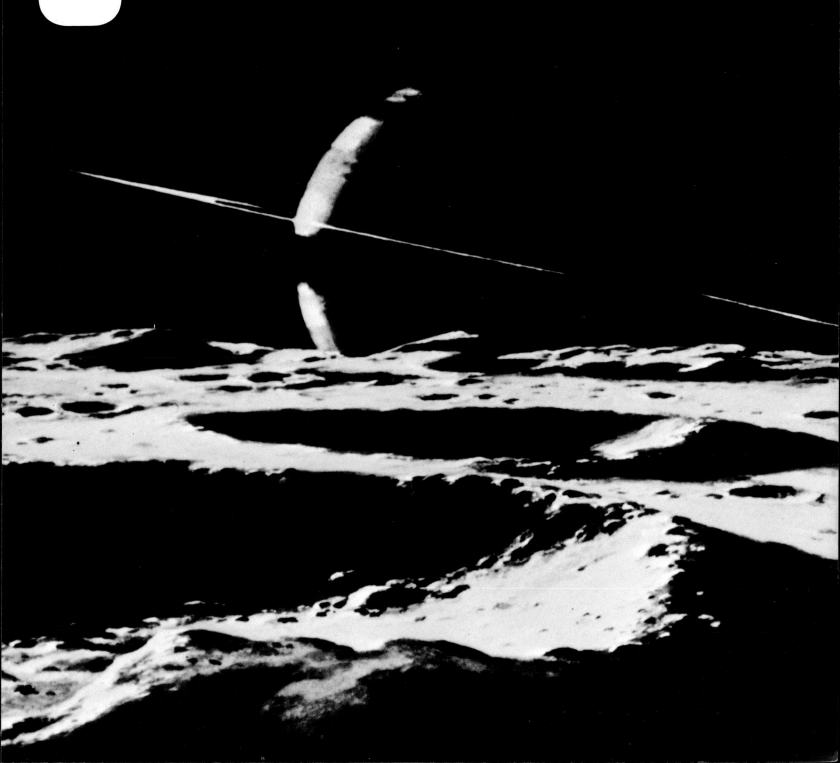

Introduction

> *Is there other life in space?*
> *Do neighbor planets have*
> *indigenous creatures?*
> *Can humans live permanently in*
> *colonies in space?*
> *Do extraterrestrials visit Earth?*
> *Can humans travel to the stars?*

Humans have been avid stargazers for thousands of years. Moderns no less than the ancients feel an overpowering sense of awe and wonder when peering into the glittering night sky. We ask some of the same questions our ancestors did. Today dramatic space photographs add to our knowledge while simultaneously increasing our curiosity.

The picture of Earth as a spherical space colony was brought home by the first photographs taken from spacecraft. Modern technology extends human vision far beyond the visible spectrum and exposes exotic worlds that excite the imagination.

This book is a unique scientific update on all aspects of life in space from the origin and destiny of human beings to possible existence of extraterrestrials. Included are a realistic assessment of current space exploration, UFO's, and projected human colonies in space. The book combines a dramatic narrative with exquisite photographs to give readers a new look at today's answers to ageless questions.

The newest look into space is truly exciting. Modern science adds daily to our knowledge of life in space while simultaneously raising tantalizing new puzzles. Exobiologists have the technology at hand to conduct a systematic search for extraterrestrials in our Milky Way galaxy *now*. Space scientists are preparing on Earth and in orbiting spacecraft for tomorrow's space treks. First contact with extraterrestrials could occur at any time. Daring human pioneers could establish the first space colony within twenty-five years.

The most thrilling discovery in scientific history could take place while *we* are alive. Whether or not there is other intelligent life in space, heroic pioneers have shown that there are no insurmountable hazards to prevent humans from living beyond Earth. The splendid views here beckon humans to venture forth, to carry life and the best of Earth into the galaxy!

Dinah L. Moché

Solar satellite power system in orbit around Earth.

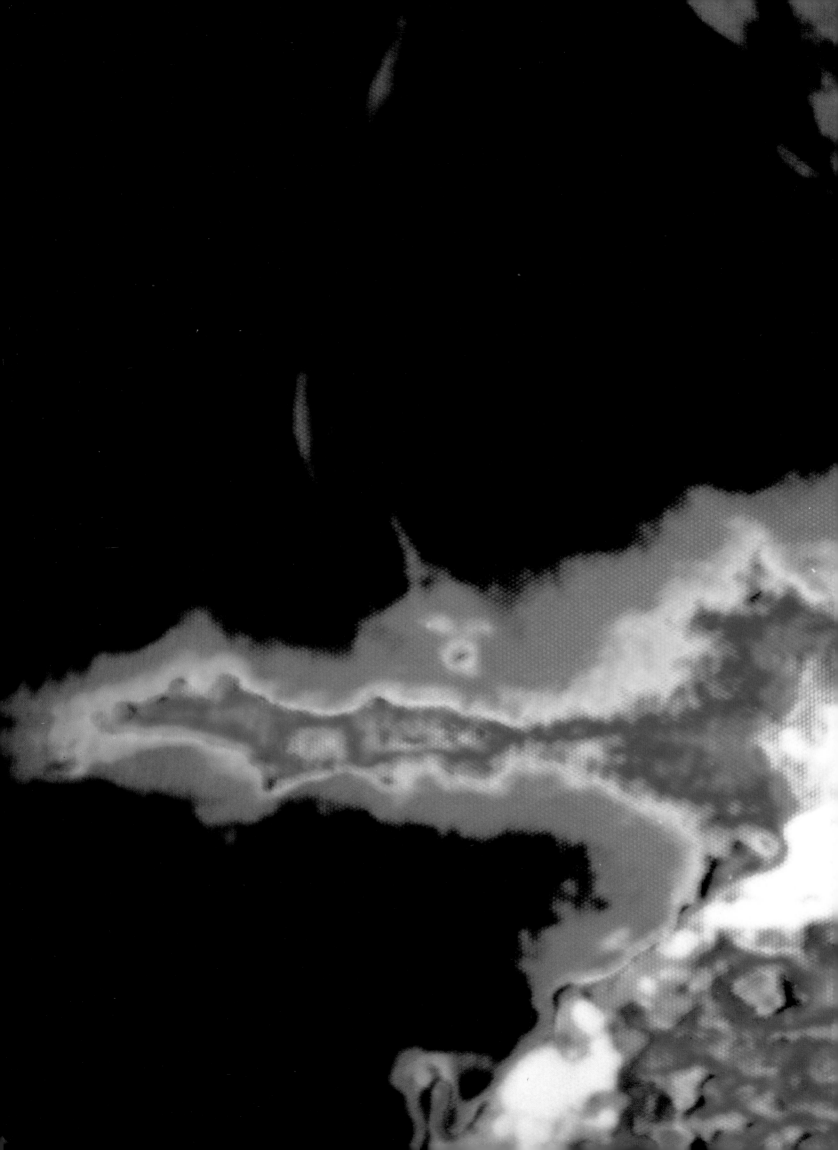

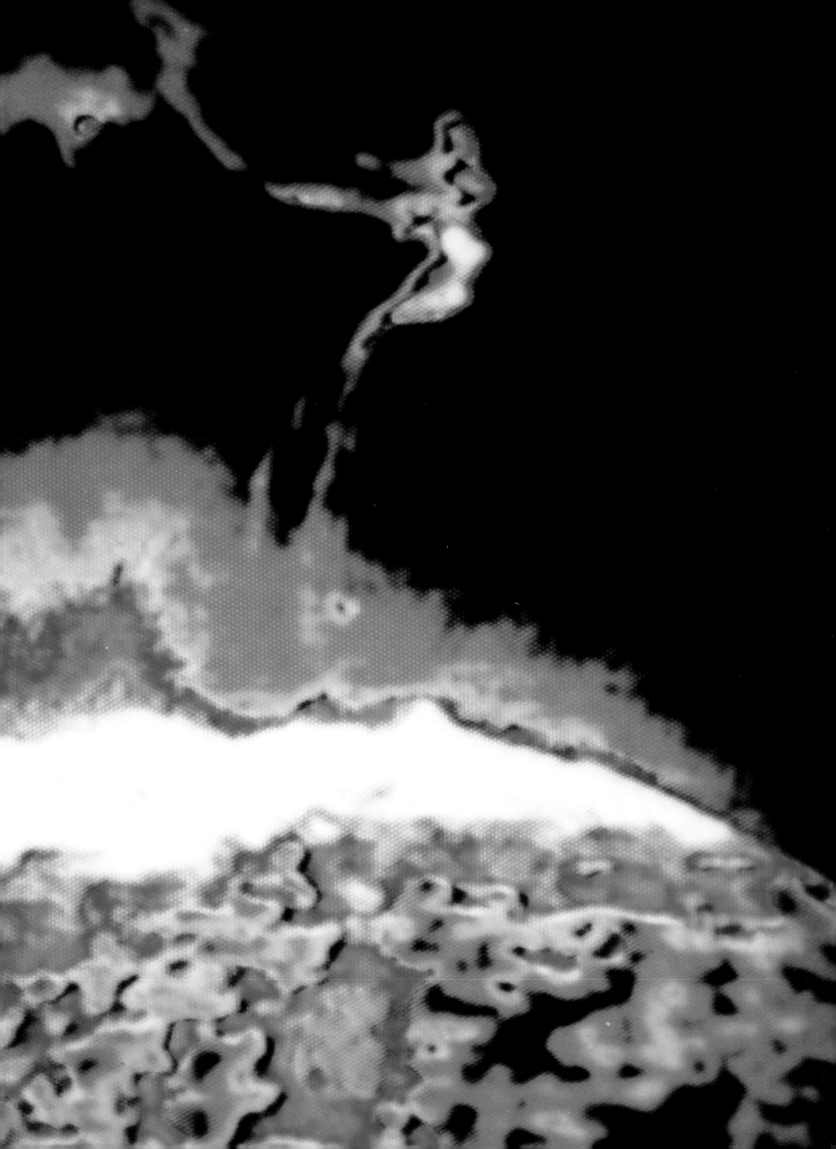

Riddle of Life

For thousands of years people have gazed in wonder at the starry sky. The ancients asked the ageless question, "Is there life in space?" Recent scientific discoveries suggest there is! Every living organism we know is made of chemical ingredients that are common in the universe. Billions of stars like our sun radiate the necessary energy. If life evolved naturally from inanimate chemicals on Earth, then it likely developed in many other parts of the universe as well.

The popular cosmic evolution theory traces our roots all the way back through the eons to the beginning of time and space. The trail through unicellular organisms, prebiotic chemistry, and the formation of planets, stars, and galaxies is incomplete and sometimes controversial. But the theory's implication is awesome. If our origins are locked into the evolution of the universe itself, we cannot be unique. The universe must be teeming with life.

The Evolving Universe

The cosmic evolution story of life starts at the moment of creation. Once the universe exploded into existence, life evolved naturally and inevitably under cosmic forces.

Our universe was born eighteen to twenty billion years ago as a titanic blast called the "big bang." In the beginning, all matter was packed into an incredibly hot dense glob that blew up violently. This primordial fireball was a cosmic hydrogen bomb. Blazing light and heat waves shot out in all directions.

When the first second of time ended, the temperature was about 18 billion degrees Fahrenheit (10 billion degrees C). The universe was filled with radiant energy. Matter in the flying debris probably consisted only of the most fundamental particles in nature—electrons, protons and neutrons. The universe has been expanding ever since the initial cataclysm. Its initial supply of energy and matter has repeatedly changed form and been recycled.

The universe cooled as the fireball expanded. Very early in time the temperature dropped to about 18 million degrees Fahrenheit (10 million degrees C). In that inferno matter zigzagged wildly at tremendous speeds. Particles frequently crashed into each other with tremendous force. Many fused upon impact, producing hydrogen and helium.

Hydrogen and helium are the most abundant chemical elements in the universe today. These gases are found in and between galaxies, stars, and planets. The very same hydrogen atoms formed in the earliest moments of time make up sixty-three percent of the atoms in our bodies. (Helium doesn't combine readily with other chemical elements.) It is not unreasonable to suppose that many other hydrogen atoms have been built into extraterrestrial creatures.

As the universe expanded further, its temperature fell too low for chemical elements heavier than helium to form by nuclear fusion. They would be produced millions of years later by stars.

Fast-flying hydrogen and helium atoms zoomed about randomly. Occasionally, gas swarms occurred that contained an unusually large number of atoms. Some swarms dispersed. Others were too

Preceding pages: Solar eruption. Theory of Cosmic Evolution suggests universe was born 18 to 20 billion years ago in gigantic explosion called "big bang." New perspectives on the creation have become possible through advanced technology, such as this color-density photograph of a solar flare taken by Skylab. Colors darken with thickness of eruption. Above: NGC 224 Great Galaxy in Andromeda, 2.2 million light-years away, as seen through modern telescope.

Nebulas are birthplace of stars, and stars are crucial link between living organisms and early universe. Orion nebula (opposite) is 1,500 light-years from Earth. Often dark patches—absorption nebulas—are seen close to bright emission nebulas. Some, like Horsehead (left), get name from unique shape. When globules of interstellar dust and gas collapse under own weight, star formation begins. Individual pieces form rotating disks. "Disk star" (above) in constellation Cygnus may be first discovery of star in process of forming planets.

massive for their gas atoms to escape. The gravitational attraction between neighbor atoms locked each in place. Gigantic collections of atoms—enough to make 100 billion suns—accumulated in roughly spherical, slowly rotating clouds.

The colossal clouds, together with free particles and radiant energy, continued racing apart with momentum generated by the primordial fireball. The universe is still expanding today. Matter is still in motion. Radiant energy left over from the big bang still testifies to the creation.

The fireball radiation has by now cooled to a temperature only 2.7 degrees above absolute zero

(−455°F, −270°C). It was discovered by physicists Arno Penzias and Robert Wilson of the Bell Laboratories in New Jersey while they were using their twenty-foot-high (6-m) horn-shaped antenna for communications satellite work. Penzias and Wilson won a 1978 Nobel Prize for discovering this faint microwave radiation striking Earth equally from every direction, just as expected eighteen to twenty billion years after the universe's beginning.

Galaxies

A few billion years after the universe began, it was filled with fiery gas and intense radiation

Top: Milky Way as it appears in summer sky. It belongs to group of 20 galaxies within a sphere three million light-years across and is spiral in form. Most galaxies seem to belong to clusters numbering from several to thousands. Globular star cluster in Canes Venatici (right) is held together by common gravity. Opposite: Constellation Taurus contains Pleiades (left)—an open cluster—and nebulas. Right: Man's imagination often credits evolutionary processes with creation of monstrous aliens, like Metaluna mutant from This Island Earth.

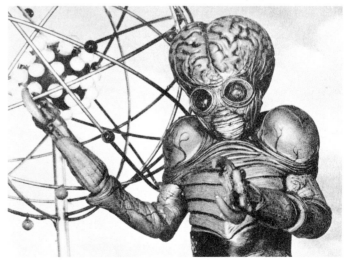

fields. The swirling mammoth clouds began to collapse slowly under their own weight. They contracted, rotated a little faster, heated up and radiated energy and collapsed again repeatedly. Finally the rotation's outward force halted the collapse caused by the force of gravity. The superclouds were then galaxies. Inside the galaxies violent changes and redistributions of the primordial matter and energy continued inexorably.

So it was that our home galaxy, the Milky Way galaxy, was formed more than ten billion years ago. By now its rotation has flattened the galaxy into an enormous disk with a faint halo of old stars in globular clusters. The disk, with its bulging central nucleus, measures 100,000 light-years in diameter and 2,500 light-years in thickness. (A light-year is almost six trillion miles, or 9.5 trillion km.) Most of the galaxy's 100 billion stars sparkle in a spiral pattern in the disk. The Earth with all its inhabitants circles the sun three-fifths of the way out from the center.

The Milky Way galaxy still rotates slowly, completing one whole turn around its center every 250 million years. Only one-fiftieth of its circuit has been completed since the first *Homo sapiens* walked the Earth. Earlier the immense times and spaces in our galaxy's past could have harbored numerous plants and animals different from those that presently exist.

There are billions of other galaxies out in space, as far as our most powerful telescopes can see. Most seem to belong to clusters that contain from several to thousands of member galaxies.

As the most massive central body, sun provides gravitational bond keeping Earth and companion planets in orbital paths. Opposite: Solar spectrum identifies elements of hydrogen, helium, and calcium in sun's composition. During solar flares (above) sun emits billows of particles and other matter of varying intensity, shown in color-enhanced spectroheliograph (right). Emissions collide with Earth's high atmosphere, resulting in forceful electrical and magnetic disturbances.

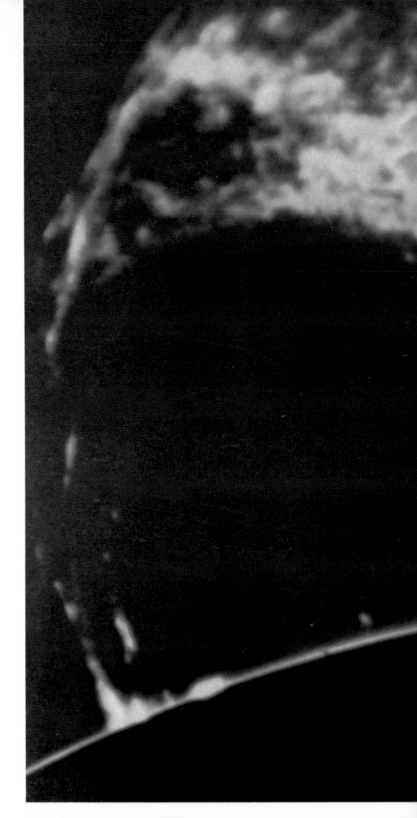

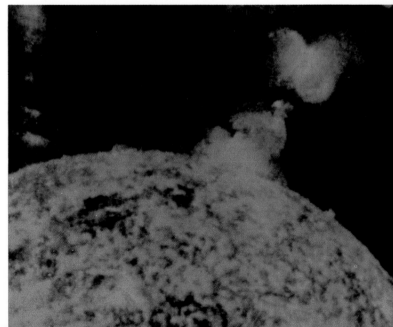

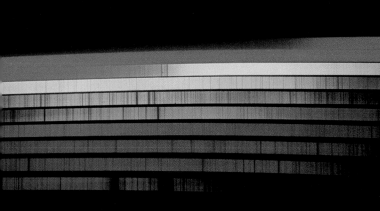

Clusters are tied together by their enormous gravity as they move around a common center of mass. Our Milky Way galaxy belongs to the Local Group of twenty galaxies located within a sphere three million light-years across. Astronomer George O. Abell of the University of California at Los Angeles has catalogued nearly 3,000 clusters containing over 1,000 members each, just within a space of four billion light-years around Earth. Farther out, there must be many thousands more.

Galaxies come in three principal patterns: spiral, elliptical, and irregular. Spirals resemble our Milky Way galaxy. Their brilliant spiral arms stretch outward from the nucleus and trail through the disk in the rotational plane of the galaxy. Ellipticals are more compact, ranging from round to flattened egg-shape in appearance. Irregulars fit no geometric form, just as their name implies.

All distant galaxies appear to be flying away from each other, still energized by the primordial fireball. Our own galaxy is hurtling through space at a speed of a million miles an hour (1.6 million km) toward the constellation Leo. All galaxies seem to be equally old. Ellipticals glow with a reddish light indicating that practically all of their stars are very old. Spirals have a bluish radiance in the great arms, indicating that hot massive stars recently formed from the plentiful gas and dust there.

A few galaxies are visible to the naked eye as faint cloudy spots in the sky. One of these, the Andromeda galaxy, actually blazes with the light of 200 billion stars. Its true size and shape are like our galaxy's. It looks deceptively tiny because it is two million light-years away from Earth. If humans have cosmic cousins in the Andromeda galaxy, they must likewise see our huge Milky Way galaxy as but a speck in eternity.

Violent changes and redistributions of the initial matter and energy in the universe can still be seen occurring in galaxies today.

The giant elliptical galaxy M87 in the Virgo cluster throws off two enormous jets of matter and energy in opposite directions. Galaxy M82 in Ursa Major appears to be exploding. Prodigious streamers of hydrogen gas burst from its center at speeds over two million miles per hour (3 million km). The nucleus of our own galaxy ejects vast amounts of radiation, high-energy particles, and massive gas clouds.

Stars

Condensations on a smaller scale inside billions of galaxies spawned myriad stars. Stars are a crucial link between living organisms and the early universe. During a cycle of changes from birth through death that takes place over millions or billions of years, stars provide the energy and most of the chemical raw materials of life.

Sunshine provides the light and heat energy for life on Earth. The sun is a second or third generation star. It is made of gases processed and recycled by earlier stars. Our very bodies are made of atoms fabricated eons ago and enormous distances away from Earth. All of the carbon in our genes, calcium in our bones, and iron in our blood were produced many billions, or even trillions, of

Space shuttle will be among first steps in ever-expanding effort to discover life on planets beyond our solar system. If nebular theory—that many planetary systems condensed from gaseous disks like one whose hot portion became our sun—is correct, then stars with planets should be common in space. In this illustration, solid rocket boosters, which augment space shuttle orbiter's main engines for initial phase of launch, are jettisoned at about 43-km altitude. Powered by solid propellant, boosters will each have about 1.1 kilograms of thrust.

years ago inside aging yet potent red giant stars.

The first stars condensed out of primordial hydrogen and helium clouds swirling inside galaxies. Hundreds of huge glowing clouds enriched in gas and dust, called emission nebulas, are still visible today inside our Milky Way galaxy. These must be the birthplace of stars. The brilliant Orion nebula is 1,600 light-years across. Because it is located 1,500 light-years from Earth, it appears to the naked eye merely as a hazy patch of light in the belt stars of the Orion constellation. Telescopic photographs using time exposure reveal its beauty.

Emission nebulas glow with light from hot newborn stars embedded inside. Some have shapes that gave rise to fanciful names. The Rosette nebula shines 3,600 light-years away in the Monoceros constellation. Measuring forty-nine light-years across, it looks like a magnificent cosmic rose.

Often dark patches are seen near the bright emission nebulas. These are absorption nebulas made of relatively dense concentrations of cosmic dust. They absorb or scatter starlight and so block our view of stars in back of the dark cloud. Some dark nebulas have also been imaginatively named from their shapes. The great Horsehead nebula is one that towers in front of a bright star field almost 1,000 light-years away from Earth.

Stars must inevitably form out of the matter in immense turbulent clouds in many parts of space. These clouds are very irregular according to the widely accepted nebular theory of star formation. They have random lumps where high-speed gas

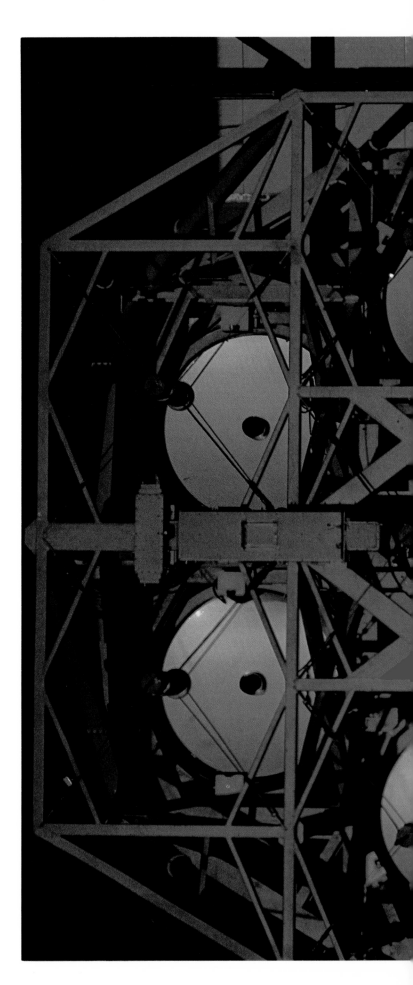

First step in detecting other planetary systems involves collecting sufficient data for analysis. Optical observatories with giant reflecting telescopes have parabolic mirrors to collect and focus starlight. Multiple Mirror Telescope (MMT) in southern Arizona synchronizes light collection by six 1.8-m mirrors on a common mount. The six perform in unison as a single, powerful 4.5-m mirror. This device is possibly the forerunner of telescopes of the future. MMT was developed by the University of Arizona and Harvard-Smithsonian Center for Astrophysics.

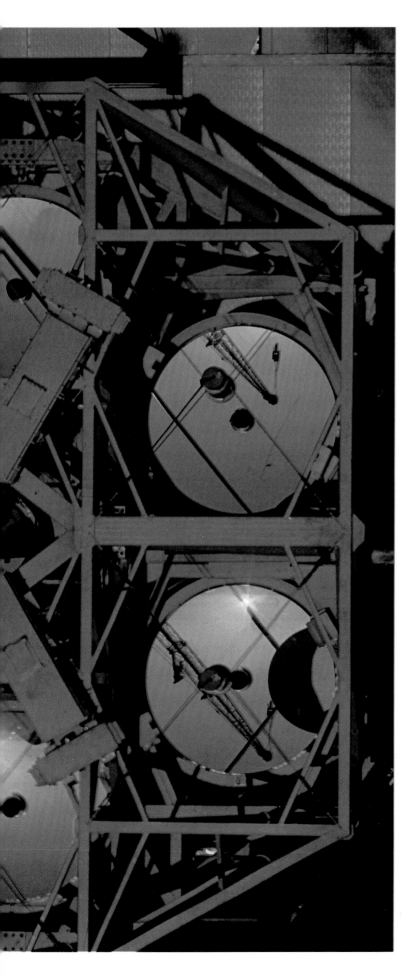

and dust particles momentarily accumulate and exert strong gravitational attraction on nearby matter. Additional gas and dust are drawn to the nucleus so the globule grows in size. Such globules can accumulate enough material for many suns. The largest may grow several times larger than our entire solar system. These are evident in front of bright nebulas.

Star formation begins when a globule gets so massive that it starts to collapse under its own weight. Forces inside the shrinking cloud mold the individual spinning fragments into the shape of a rotating disk. As the disk contracts under gravity its density increases. The internal temperature and pressure soar. Heat flows from the hot center to the cooler surface. The new body radiates this energy into space. It glows with a deep red color.

Within a few million years the internal temperature rises to eighteen million degrees Fahrenheit (10 million degrees C). The core bursts into the flame of nuclear fusion and a new star is born! Deep in the fiery turbulence of the star's core groups of four hydrogen nuclei are fused into single lighter helium nuclei. Each time, the mass difference is released as energy, according to Albert Einstein's famous equation $E=mc^2$ where E=energy released, m=mass lost, and c is the speed of light.

Energy generated by nuclear fusion reactions in the core ultimately is radiated out into space. The very high internal temperatures and pressures are maintained by the thermonuclear reactions. The outward pressure of the very hot gases balances the inward pull of gravity. The star shines its own light steadily into space for ongoing ages.

Our sun was formed this way about five billion years ago. Its birth process took over thirty million years. Stars ten times more massive than the sun require only a few hundred thousand years to light up. Those with a mere one-tenth our sun's mass need about 100 million years.

Our sun has changed little since its birth. it is not expected to change noticeably for another five billion years. Stars spend 99 percent of their active lives burning hydrogen and radiating energy into space. During that time they supply vast quantities of energy that may support life.

When its hydrogen fuel is finally depleted, a star begins to die. Then dramatic changes take place in its structure and appearance. It swells to gigantic proportions and shines red light into space. Because of its color and size, the star is then called a red giant. When the sun becomes a red giant after another five billion years, it is expected to engulf the planets Mercury, Venus, Earth, and Mars. Hopefully, humans will have found new homes in space long before the sun dies.

Ultimately all stars go out. Stars like the sun throw off a gas envelope called a planetary nebula. Then they shrink under their own weight to a super-dense white dwarf star no bigger than Earth. The wispy planetary nebula floats atoms away for recycling. The white dwarf gradually shines its last light into space and becomes dead forever.

Red giant stars much more massive than the sun are the fiery sites where chemical elements heavier than hydrogen and helium are fused. Spasmodic death contractions and expansions generate a series of rises in the star's core temperature, pressure, and density. Ongoing nuclear fusion reactions in the core produce such important elements of life as carbon, oxygen, nitrogen, and calcium. Finally, the fusion of iron quenches the nuclear fires.

Very massive stars die with a brilliant violent explosion called a supernova. They leave exotic stellar corpses and hurl chemically enriched debris back into space. Their final inferno spawns the heaviest chemical elements. Supernovas spray the fabricated chemical elements around their galaxies for recycling into stars and rocky planets.

The sun and Earth are typical of such recycled objects. Fifteen billion years of cosmic events had occurred before the sun and Earth even started taking shape. Millions of very massive stars had already lived full lives and died, enriching our part of space with a panoply of chemical elements. Every atom in every plant and animal on Earth was created long ago and far away in natural universal processes. Every living thing on Earth is literally made of star dust!

Planets

Extraterrestrials could easily look quite different from earthlings, but they would most likely inhabit a suitable planet circling a sun. They certainly can't exist in a blazing star. Planets are much smaller than stars and don't emit as powerful radiations. Although astronomers cannot yet actually see planets beyond our own solar system, most agree that myriad stars command planetary systems.

Our sun is just one of 100 billion stars in the Milky Way galaxy. There are some 100 billion galaxies in the universe. Thus, there are at least 10,000 billion billion stars in space. Even if the chance of a star having a planet like Earth were only one in a million, there could be 10,000 trillion other planets with living creatures!

According to the nebular theory, our planetary system condensed from the same huge rotating gaseous disk whose dense, hot central portion became the sun. Irregular lumps of matter that formed randomly away from the center of the turbulent cloud gravitationally attracted more material that accreted into planets. The sun's nine planets are still circling it in the same direction and plane.

Today the sun's atoms are 87 percent hydrogen, 13 percent helium, and less than 1 percent oxygen and nitrogen. The nebular theory accounts for the current differences in composition between the sun and its planets. Hydrogen and helium from the outer regions of the parent disk are evident in the dense atmospheres of the giant planets Jupiter and Saturn. Planets such as Earth, which are closer to the sun, could have lost their original hydrogen and helium atmospheres in blasts of solar wind from the young sun. They could have formed without atmospheres at all. Heavier elements were locked inside these planets. Volcanoes could have released different gases and water from the planets' interiors early in their history.

Apollo 11 astronauts Neil A. Armstrong, Edwin E. Aldrin, Jr., and Michael Collins were launched to moon by Saturn V launch vehicle in July, 1969, from Cape Kennedy. Left: Aldrin deploys Early Apollo Science Experiments Package and Passive Seismic Experiments Package on lunar surface. Observations of other planets by actual sample is most effective and accurate way to collect data. After take-off from moon, Armstrong and Aldrin joined Collins in command module that was circling the moon during mission.

If the nebular theory is correct for the sun's planetary system, then stars with planets should be common in space. Unlike catastrophic theories of planet formation that rely on events such as singular explosions or collisions, the nebular theory implies that whenever stars form, planets will likely form as well. The sun and Earth were not cosmic accidents. Rather, the universe is probably full of potentially life-supporting planets.

Astronomers Peter Strittmatter and Rodger I. Thompson of the University of Arizona recently photographed a star—MWC 349—that radiates an extraordinary amount of visible light. This, together with infrared data, suggests that MWC 349 is the first planetary system in formation ever seen.

Chemical Evolution

All planets and animals on Earth are made of complicated chains of molecules built around the chemical element carbon. These carbon-based molecules are called organic for their important role in life processes. Two well-known examples are deoxyribonucleic acid (DNA) and ribonucleic acid (RNA), which control the mechanisms of heredity.

Carbon is one of the more abundant elements in the universe even though it is not nearly as plentiful as hydrogen and helium. Another reasonably abundant element that can link atoms simultaneously into complex chains is silicon. Furthermore, radio astronomers have detected in interstellar clouds more than forty different molecules, including both water and organic compounds that are crucial to all known plants and animals. The raw materials of life exist in many parts of our galaxy and in other galaxies, too!

Earth's environment five billion years ago was probably very different from the world today. The first stable atmosphere likely resulted from outgassing by volcanic action. Volcanoes eject water vapor, carbon dioxide, nitrogen, and other gases together with lava. These could have added hydrogen, methane, and ammonia to the air by chemical interaction with rocks. Chemical evolution—a series of naturally energized chemical reactions—could have led from hydrogen, nitrogen, methane, ammonia, carbon dioxide, and water to living cells billions of years ago.

Since the first dramatic laboratory experiment in 1953 by Stanley Miller and Harold Urey, many experiments have repeatedly shown that the simple gases presumed to be present in Earth's early atmosphere can be energized into prebiotic

Simulation of in-space experiences are necessary if manned flights are to achieve maximum success. Above: Viking's flight was simulated in 1971 in preparation for 1975 launch. Opposite: Voyager I's cameras show Jupiter's moons (top) Ganymede (right, center) and Europa (top right), in 1979. Despite small images, in-flight photos show details of satellites not seen in photos from Earth. Gemini IV manned space flight allowed astronaut Edward White first extravehicular activities and first use of personal propulsion unit (right).

compounds, those that exist before life. Ultraviolet light from the sun, lightning, ionizing radiation, and heat energy were all available to activate the synthesis of simple organic molecules. These could have subsequently linked into molecules of greater complexity.

Early Life

Organic molecules are the units from which all Earth life is built, but they are not alive. What caused the spark of life? Did the first living cells develop inevitably in Earth's primitive environment or were they a singular cosmic occurrence? No one knows, but the cosmic evolution theory neatly ties the appearance of unicellular organisms to universal forces.

Water speeds up chemical reactions. Over four billion years ago organic matter accumulated in Earth's seas. Collisions between randomly moving molecules occurred frequently in the water. Violent impacts joined small molecules into larger ones. The concentration of molecules increased in shallow seas and in lakes where water evaporated continuously.

Increasingly complex molecules formed over billions of years. Eventually molecules evolved that were capable of reproduction. In this way the threshold from inanimate molecules to living matter was crossed naturally and inevitably. Simple one-celled plants called algae and organisms such as bacteria are found as fossils in rocks over three billion years old. Unfortunately, no geological records have survived to picture the events from the time the molec-

ular precursors of life existed in Earth's seas up to the time the first living organisms appeared.

Organic matter seems to have evolved elsewhere in our solar system. Some meteorites, called carbonaceous chondrites, contain up to 5 percent biologically significant compounds. These same compounds are suspected to exist in the atmospheres of the outer planets and some of their moons. Perhaps primitive organisms exist there, too.

Ascent of Humans

All of the world's diverse creatures evolved from the first unicellular organisms over the next three billion years, most biologists agree. Biological evolution depends on mutations for modification of genetic material. Favorable traits are preferentially retained in the gene pool. They give an organism a greater chance of survival in competitions for energy sources and in coping with environmental stresses. As generations pass by, species with favorable mutations will slowly but inevitably replace those without them. This process of natural selection explains how the many different plants and animals we know evolved.

Homo sapiens appeared only very recently in the evolutionary chain. Intelligence has been a favorable trait so far. A sophisticated scientific and technological civilization has emerged. The infinitesimal fraction of cosmic atoms that are packed into the human brain make thought possible. And so with our primordial brain atoms, we speculate that other intelligent creatures have evolved beyond Earth and we ponder greeting them.

New Frontier in Space: Self-sufficient colonies (top), each capable of housing, feeding, employing, and entertaining 10,000 people, are established amid splendors of Milky Way with mother Earth appearing as moon in background. Bottom: Fanciful concept by artist Edward A. Anderson of extraterrestrial polar community, location unspecified. This planet appears to have atmosphere permitting moderate, Earthlike climate. Plus liquid water—presently unknown in space. Existence of water gives hope of discovery of life.

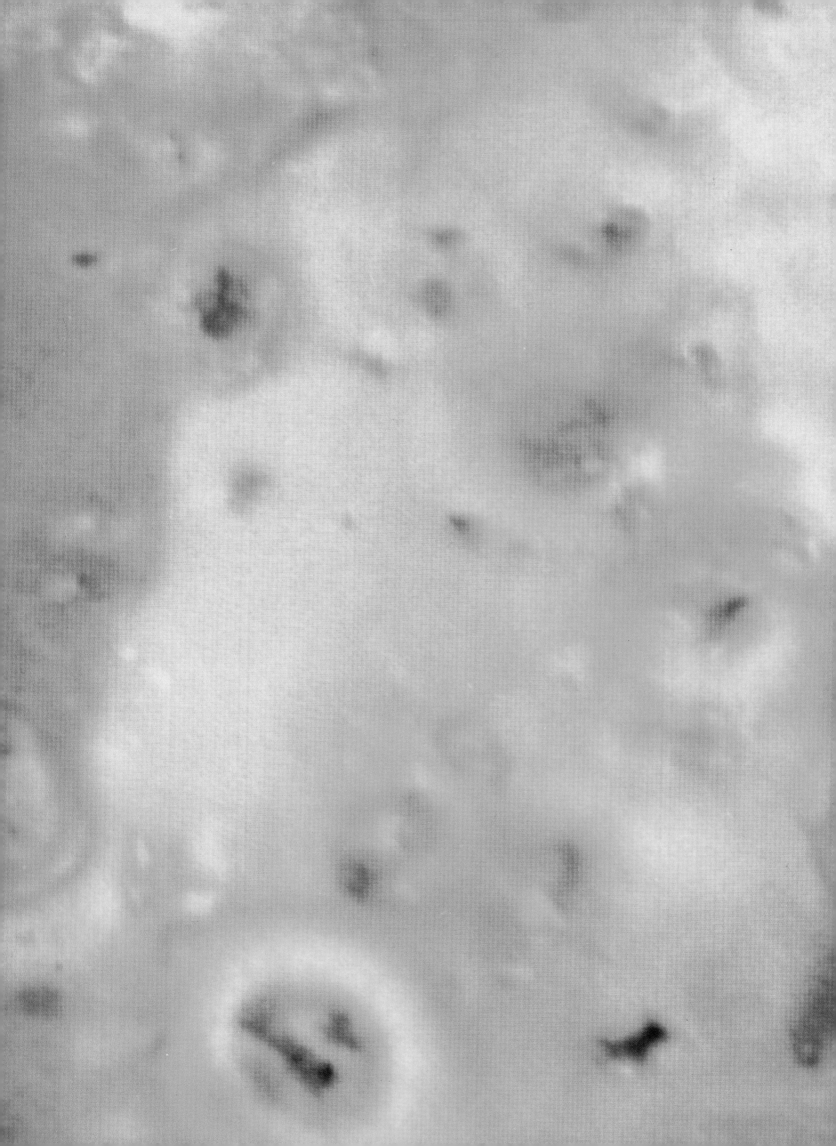

Planet Close-Ups

Simple organisms might have evolved on other worlds in our own solar system. Sophisticated robot spacecraft have been sent to other planets to determine local conditions and radio their findings back to Earth. Their close-ups suggest that life can exist on neighbor worlds! Today that life might be native microbes. Tomorrow it could be humans.

Mars

Of the eight other planets in our solar system, Mars is the most likely to have native inhabitants. Located an average 142 million miles (227 million km) out from the sun, the planet receives sufficient sunlight and heat to support life. There is strong evidence that life could have evolved from the raw materials and water there.

Mars is best observed at 780-day intervals when its orbit brings it closest to Earth. Then, features that have long excited speculation about colorful Martians are easily seen in a small telescope. A white cap spreads over the north and south poles when it is winter in each hemisphere. These polar caps shrink in the spring and summer seasons. Sprawling dark regions that sweep over the globe in the Martian spring look deceptively like blooming vegetation or water spilling from melting polar ice. Illusory streaks even suggested canals that romantic observers at the beginning of this century thought might have been constructed by intelligent Martians.

It is understandable that until recently visionaries populated Mars with exotic little humanoids. With a diameter of 4,220 miles (6,750 km), Mars is half as big as Earth. Because of its small size and mass, the planet has weaker gravity. A 200-pound (90-kg) person would weigh only seventy-six pounds (34 kg) on Mars. The Martian year has seasons that resemble Earth's, although it lasts 687 days. A day lasts twenty-four hours and thirty-seven minutes.

American and Soviet robot spacecraft began to fly by and photograph Mars in 1965. Space-age discoveries eliminated the fanciful little green canal-builders of earlier imaginings. But they do hint at simple Martian organisms!

Mariner 9 orbited Mars in 1971 and sent back pictures of huge, winding channels that look as if they were carved out by mighty floods eons ago. Mars is currently in an ice age. It is too cold for water to flow, but the dry channels with tributaries look exactly like those eroded by great rivers on Earth. Perhaps they were sculpted during a warmer age in the planet's history by water that is now locked in subsurface permafrost and polar ice caps.

If Mars did once have a warmer climate and large amounts of flowing water, life could have evolved there as it did on Earth. If life did develop, it could have survived by natural selection as environmental conditions worsened over the millennia. Martian organisms would have to be specially adapted to live on that nearly dry and airless planet, but there is no apparent reason why they could not be. Finding life of any kind on Mars would greatly increase the probability that life exists elsewhere in the universe.

Two spacecraft, called Viking 1 and 2, arrived at Mars in 1976 equipped to study its atmo-

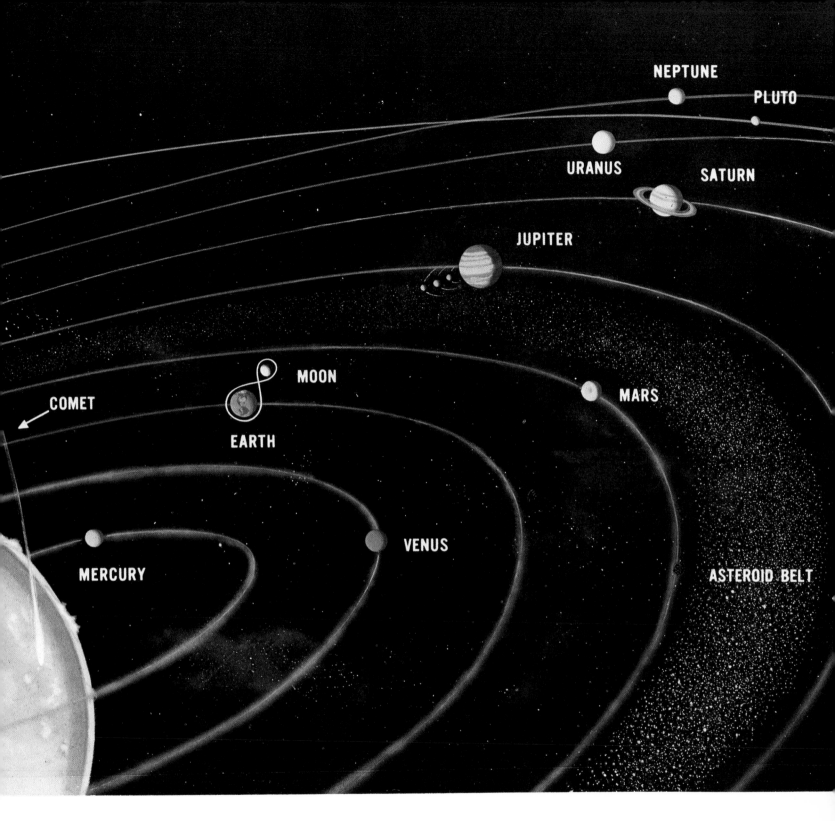

NEPTUNE

PLUTO

URANUS

SATURN

JUPITER

MOON

MARS

COMET

EARTH

MERCURY

VENUS

ASTEROID BELT

Preceding pages: Blotchy complexion of Io, seen from 377,000 km, probably is a mix of surface deposits. Voyager 1's extraordinary discoveries of widely varying compositions of Jupiter's Galilean moons increase possibility that preconditions for life exist within our solar system. Above: For first time mankind is getting closeup views of its planetary neighbors. Voyagers 1 and 2 will cross orbits of Saturn, Uranus, Neptune, and Pluto before entering interstellar space.

sphere and geology and to search for life at two ground sites. Each Viking was actually two robot explorers. An orbiter photographed the surface of Mars and analyzed the atmosphere while circling hundreds of miles over the planet. A lander parachuted to probe the ground.

The Vikings were thoroughly sterilized before leaving Earth to reduce chances (below one in a million) that they would take our microorganisms on their search for extraterrestrial life. On July 20, 1976, eleven months and 400 million miles (640 million km) after launch, the Viking 1 lander separated from its orbiter. Its descent was automatically controlled by an on-board computer because commands from Earth took twenty minutes to reach Mars. Viking 1 settled into powdery dirt in a plain called Chryse Planitia (Plains of Gold) located at 22.46°N latitude, 48.01°W longitude. On September 4, the Viking 2 lander set down in Utopia Planitia (Plain of Utopia),

located about 4,600 miles (7,400 km) northwest of its predecessor at 47.89°N, 225.86°W.

The Viking landers were admirably capable robots. Each had two camera eyes for scanning almost a full circle, a long arm with a small shovel to scoop soil for tests, the equivalent of two power stations for energy, and two computer-center brains. In addition, each packed a weather station, an earthquake detector, two chemical laboratories, and three biology laboratories in their small car-size bodies.

The landers never left their landing sites. Instead each stored data on magnetic tape as it made discoveries. Bits of information were periodically transmitted via the orbiters to eager scientists on Earth. Data was processed by computers into spectacular pictures.

The landing sites are bright red. Rocks are scattered among fine-grained material that looks

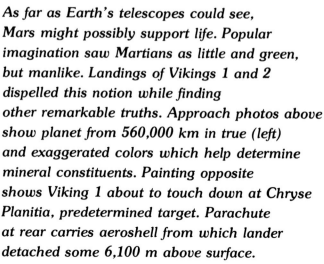

As far as Earth's telescopes could see, Mars might possibly support life. Popular imagination saw Martians as little and green, but manlike. Landings of Vikings 1 and 2 dispelled this notion while finding other remarkable truths. Approach photos above show planet from 560,000 km in true (left) and exaggerated colors which help determine mineral constituents. Painting opposite shows Viking 1 about to touch down at Chryse Planitia, predetermined target. Parachute at rear carries aeroshell from which lander detached some 6,100 m above surface.

like iron-rich clay. A reddish covering on the rocks looks like an iron oxide compound from chemical weathering. The sky is colored pink in the daytime by red dust in the atmosphere. Sunsets are pale blue.

No Martians showed up in front of the cameras. But they might have revealed themselves in the landers' biology experiments. The Vikings searched for microbes in the soil. Each robot scooped a bit of Martian soil with its shovel and dropped it into a biology instrument for tests. This sophisticated instrument was the equivalent of three biology laboratories with technicians. It had food and water, tiny ovens, lamps to imitate Martian sunlight, bottled radioactive gases, and Geiger counters all packed into a space the size of a typewriter case. The tests were based on life processes common on Earth. Here plants and animals ingest food and gases, use them to grow, and give off waste products. Experiments operated by remote control from Earth sought these life signs on Mars until May, 1977.

One test, called the labeled-release experiment, produced results that suggest living organisms to many scientists. A pinch of Martian soil in a test cell was slightly moistened by a nutrient nicknamed "chicken soup" containing radioactive carbon 14. Presumably, if any microorganisms lived in the soil they would consume the labeled food, assimilate it, and release waste gases containing the telltale radioactive carbon 14 as a result of metabolic activity. A waiting radiation detector reported that radioactive carbon dioxide gas was produced in the pinch of soil. The carbon dioxide could be

Rugged Martian surface as seen by Viking 1 (top & left) and 2 (far left), some 7,400 km apart. Landscape everywhere is bright red and strewn with rocks. Top: Late afternoon shot toward horizon 3 km distant. White arm is meteorology boom. Far left: Circular structure is high-gain antenna beamed to Earth. Left: Sampler scoop and magnet cleaning brush. Vikings were stationary, but had weather station, computers, chemical and biological labs, and data transmitters aboard. Both were sterilized to prevent contamination of Martian atmosphere.

Mars features rendered from photographs taken by orbiting Mariner 9. Top left: Huge extinct volcano, Mount Olympus, measuring 603 km across base and 24 km in height. Left: Detail of collapsed center. Above: Portion of 4,800-km rift valley. Arizona's Grand Canyon would fit in a small tributary. While observations were geologically accurate, Mariner did not catch Mars' distinctive red color. Far left: Viking 2 photographs Martian sunrise on 631st day after landing. Aside from esthetics, colors aided analysis of atmospheric dust and ice particles.

due to living Martian microbes. The result can also be attributed to unfamiliar chemical reactions.

Different tests sought critical organic molecules that should be in the Martian soil if familiar microbes live there. The landers did not find any organic molecules. However, there is no reason that Martian microbes must be identical to those we know.

At present no one can say positively whether or not there is life on Mars. Scientists are continuing laboratory experiments on Earth in search of answers to puzzles raised by the lander data.

The Viking orbiter cameras and remote-sensing instruments operated for more than a year. They confirmed earlier observations that suggest that life could have developed in a warmer, wetter age on Mars.

There is direct evidence of water on the planet. Some water is frozen in the permanent ice cap at the north pole. There are fog and occasional filmy clouds. There is indirect evidence of ancient catastrophic flooding. Great and small channels resembling river systems on Earth are prominent, and some even have teardrop islands characteristic of midstream obstacles.

Orbiter photographs show a rugged globe. Several volcanoes reveal that Mars has had massive volcanic eruptions in the past. The largest volcano is Mount Olympus, which towers seventeen miles. (27 km). The huge caldera, or crater, at its summit is about fifty miles (80 km) across and almost two miles (3 km) deep.

There are vast canyons. Valles Marineris is 4,000 miles (6,400 km) long, as much as three miles (5 km) deep, and 150 miles (240 km) wide. This canyon runs along the equatorial region. It is blanketed with fine-grained material.

Wild dust storms often rage, although they did not seem to be evident from the landers. The dark areas misinterpreted by earlier speculators are probably the result of global dust storms that start in the spring. Winds as high as hundreds of miles an hour can blow light-colored dust about and expose dark surface.

The Martian atmosphere is about 95 percent carbon dioxide. It has the critical elements of life, with 2 to 3 percent nitrogen, 1 to 2 percent argon, 0.1 to 0.4 percent oxygen, and traces of water vapor and other gases. The atmosphere is too thin to block the deadly ultraviolet rays from the sun, so they strike the surface mercilessly.

Mars is very cold. The highest temperatures at the Viking 1 landing site on summer days hovered around $-22°F$ $(-30°C)$. Winter temperatures dropped lower than any ever recorded on Earth, below $-190°F$ $(-123°C)$.

Biologists E. Imre Friedmann and Roseli Ocampo-Friedmann of Florida State University at Tallahassee have suggested a way that Martian organisms could use natural shields against the deadly ultraviolet rays and severe cold. They found living bacteria, algae, and fungi in niches in rocks in the cold, arid valleys of Antarctica. Apparently the rocks trap heat and sunlight and provide a protective environment for life. The Viking landers were not designed to reach into the interiors of the Martian

Probability of finding life forms on Mars is so high that new robot missions have been proposed for 1980's. Pairs of rovers (Vikings on wheels) capable of mutual aid can be built with present technology. Here a pair execute daily traverse of up to 5 km, beam findings to mother orbiter for relay to Earth. (Second pair is over horizon.) More ambitious would be landing of robot spacecraft which could return to Earth with soil samples for analysis. Planet is cold, harsh, dry, with atmosphere 95 percent CO_2, but could be colonized by humans.

rocks to seek life, but life could exist there.

The probability of finding life on Mars is so high that scientists have proposed new robot missions for the 1980's. Pairs of roving landers capable of assisting each other could be built with present technology. They would go to Mars with cameras and instruments to probe especially interesting sites. A more ambitious mission would land robot spacecraft on Mars, collect a bit of soil, and return it to Earth for analysis.

In the twenty-first century Mars could be colonized by humans. It is a harsh, cold, dry planet but not impossible. There is adequate sunlight to supply energy for plants, animals, and people. Water and raw materials are available.

Scientific bases where astronauts, protected from the deadly environment by spacesuits, could

Armada to Mars: By 1990's advance party could be preparing way for colonization of Mars. In painting at right, square solar sails serve as Interplanetary Shuttles, have already landed two manned capsules, plus rovers and processing vehicles to mine and study Martian surface. Nuclear power station would precede astronauts, although planet has adequate sunlight for energy too. Ultimately, underground cities with suitable environments could be built. Above: Limber robot rover seizes rock sample from steep-sided arroyo under radio direction.

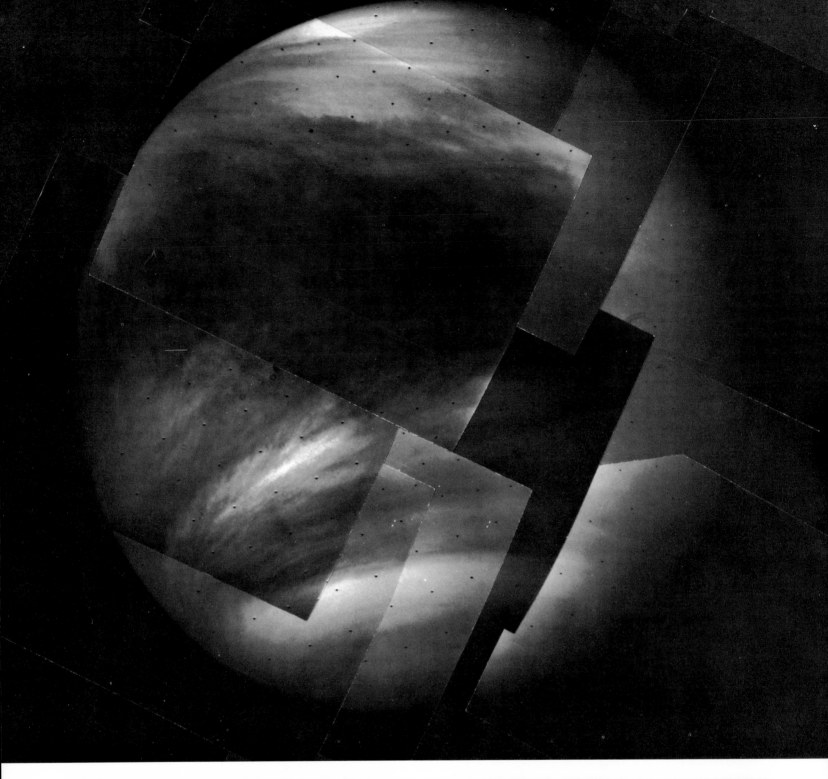

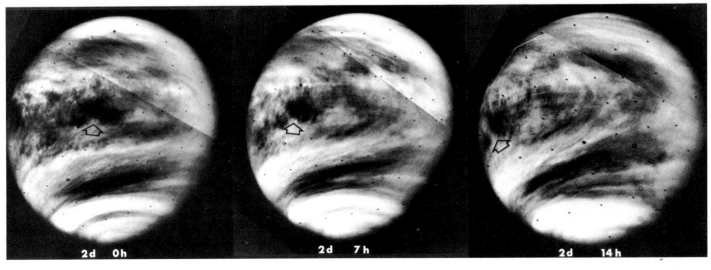

2d   0h   2d   7h   2d   14h

do experiments would come first. Ultimately, domed colonies and underground cities with suitable environments might be built where humans could live entire lives without ever visiting Earth.

The two small moons that circle Mars are not proposed landing sites, but could serve as mineral lodes for future colonists. As seen by the Vikings, Phobos and Deimos are rock chunks about fourteen and nine miles (22 and 14 km) wide respectively. Both are pocked by many craters. Phobos has long, unexplained parallel grooves. It orbits Mars every seven hours and thirty-nine minutes. Deimos, which looks smoother and appears to be covered with fine-grained soil, completes a circuit in about thirty hours.

Venus

Of all the planets, Venus is the most like Earth in size, mass, and distance from the sun. It is 7,600 miles (12,200 km) across, has mass enough to hold on to an atmosphere, and is bathed in life-giving sunshine. Despite these similarities, there is virtually no chance that creatures like us live on Venus.

The second planet from the sun, Venus is some 67 million miles (108 million km) out. It orbits the sun in 225 days, coming to within twenty-six million miles (42 million km) of Earth at times. Venus rotates extraordinarily slowly, completing a turn once every 243 Earth days. If humans ever visit Venus they will find that the days seem endless. Since the planet rotates on its axis in a direction opposite that of its orbital motion (retrograde), the time from sunrise to sunset is 117 Earth-days long.

Venus outshines all the stars in our night sky. Dubbed the "morning star" and "evening star," it is dazzling because it is shrouded in thick clouds that reflect a lot of sunlight. The clouds effectively kept Venus a world of mystery for centuries. Dreamers concocted wild tales of civilizations thriving beneath the shroud.

Space-age radar and robot probes have shattered these fantasies. The first three American Mariner and eight Soviet Venera spacecraft to Venus flew above Venus' cloud tops. They gathered data from a distance, using remote-sensing equipment. Venera 9 landed on October 22, 1975, at 33°N latitude, 293° longitude in a dry, rocky site with little soil. Three days later, Venera 10 reached the ground, some 1,200 miles (1,900 km) away at 15°N latitude, 295° longitude in an ancient plateau with weathered rocks and relatively dark, fine-grained soil. They met hellish conditions but sent back our first surface pictures before they expired.

The surface temperature on Venus is a sizzling 900°F (482°C), hot enough to melt lead. That searing heat would break up every molecule in familiar living cells. Apparently a runaway greenhouse effect creates the inferno. The planet absorbs incoming sunlight and reradiates it as heat. The heat cannot escape through the thick atmosphere. The trapped radiation heats the planet to higher and higher temperatures.

Human lungs would be poisoned and crushed by the Venusian atmosphere. It contains over 95 percent carbon dioxide gas. Air pressure is over ninety times greater than normal. It is equivalent to

Mysterious World of Venus: Shrouded in thick clouds that prevent incoming sunlight from escaping as heat, planet suffers murderously high temperatures and pressures. Though most like Earth in size, mass, and distance from Sun, Venus is extremely hostile to human beings. Top: Mosaic of Venus taken by Mariner 10 from 845,000 km. Cloud patterns, seen only in ultraviolet light, show general circulation of planet's upper atmosphere. Bottom: Time-sequence pictures show rotation of dark marking atop Venus cloud cover. Feature measures about 1,000 km across.

going down in the ocean. At present exploration of Venus must be by robot craft suitably built to survive its murderously high temperatures and pressures.

The first American spacecraft to orbit Venus arrived there December 5, 1978. Pioneer Venus 1 weighs 1,280 pounds (581 kg) with about 100 pounds (45 kg) of remote-sensing devices, control computers, and communications equipment designed to probe the planet for at least one Venusian year. It flew some 300 million miles (480 million km) to the planet, traveling outside Earth's orbit for three months after launch and then inside Earth's orbit for four months.

Pioneer Venus 1 dipped within ninety miles (144 km) of the surface to sample the composition of the upper atmosphere. From orbit it made radar measurements of the terrain and snapped daily ultraviolet and infrared pictures of the thick clouds. Data was radioed to Earth, where computers processed it and generated brilliant false-color pictures coded with information.

A revolutionary spacecraft called Pioneer Venus 2 arrived at the planet on December 9, 1978. At a distance of 7.8 million miles (12.5 million km), the spacecraft divided into five individual probes. All five capsules dove through the clouds at points 6,000 (9,600 km) miles apart. Each beamed back information about the winds, clouds, and atmosphere before burning.

Instruments inside Pioneer Venus 2 were protected by special heat-resistant aluminized-plastic sheets and lightweight titanium shells able to resist 100 times normal air pressure. A thirteen-carat dia-

mond window the size of two pennies stacked together let radiation into waiting sensors. One tough capsule unexpectedly survived impact at twenty-two miles (35 km) per hour and sent back data for sixty-seven minutes from the ground before succumbing.

The topography of Venus could be similar to Earth's. Radar images portray high mountains and extensive relatively flat areas. The planet looks very dry. It has a sensible atmosphere out to about 155 miles (249 km).

The bulky clouds are about twelve miles (19 km) thick with three distinct layers. They are located between thirty-one (50 km) and forty-three miles (69 km) above the surface. Apparently they are composed mainly of sulfuric acid droplets and microscopic sulfur particles that drift slowly down to the hotter, lower altitudes. At about twenty-nine miles (46 km) above ground the droplets apparently split up, forming water vapor, sulfur dioxide, and molecular oxygen.

The atmosphere seems to consist of about 97 percent carbon dioxide with about 1 to 2 percent nitrogen, 0.1 to 0.3 percent water vapor, and traces of oxygen, helium, neon, and argon. The amount of nitrogen is comparable to that in our air. Both planets have roughly equal amounts of carbon dioxide, too, but most of Earth's is locked in carbonate rocks. Venus may originally have had abundant water that circulated to the top of the atmosphere and was split into hydrogen and oxygen by solar ultraviolet radiation. The light hydrogen could have escaped into space and the heavier

Analyzing Venus: Ultraviolet spectrometer measures atomic oxygen (crescent) and hydrogen (bars) in Venusian atmosphere (top left). Other three images were constructed by Infrared Radiometer aboard Pioneer 10. Top right shows intensity of thermal emission, bottom left is cloud map, bottom right shows three atmospheric levels of polar region. Computer-processed data gives science three-dimensional view of Venus' meteorology. Right: Painting of Pioneer Venus Orbiter in sight of planet with all systems go. Pioneer 2 hit surface and succumbed.

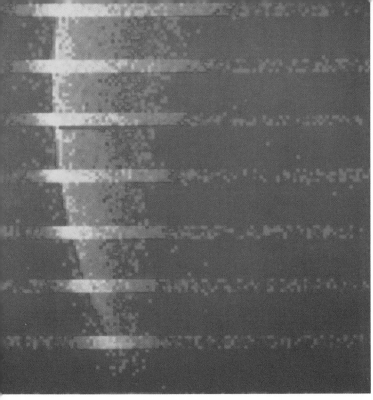

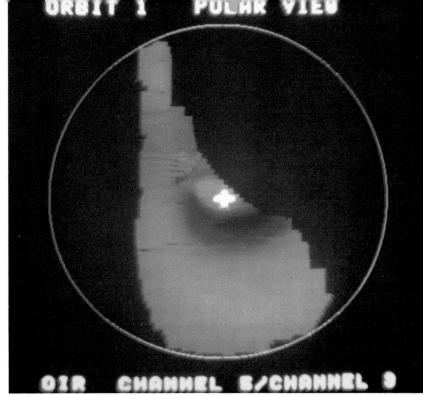

OIR   CHANNEL 5/CHANNEL 9

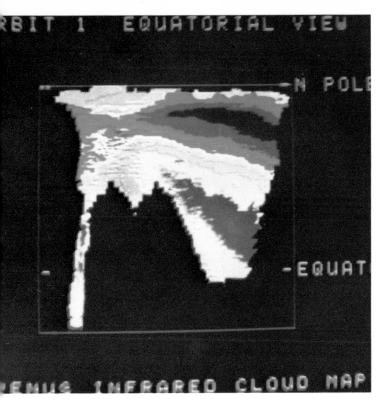

N POLE

-EQUAT

VENUS INFRARED CLOUD MAP

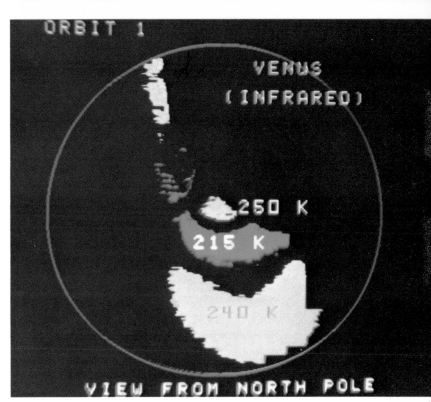

VENUS
(INFRARED)

250 K

215 K

240 K

VIEW FROM NORTH POLE

oxygen could be locked up in the crustal rocks.

Venus is not a likely place for life to have arisen, but the possibility cannot be ruled out entirely. Suppose simple life developed long ago. Through natural selection it could have evolved to rapidly reproducing forms that floated in the atmosphere between the hellish surface and the acid clouds. Such floating microbes would have to be hardy indeed to survive today.

Venus continues to fascinate explorers. The Soviet Venera 12 arrived there December 20, 1978, and an identical companion, Venera 11, followed on Christmas Day. Each Venera separated into two robots. Landers carrying scientific equipment parachuted to the torrid surface about 500 miles (800

Above: Torrid surface of Venus as seen by Soviet Union's Venera spacecraft (right). Venera 11 and 12 landers parachuted in, 800 km apart, and managed to transmit data for 95 and 110 minutes respectively to hovering space stations for relay to Earth. Venera 12 reported temperature of 460°C and air pressure 88 times Earth's. Both saw lightning flashes and generally red light since only Sun's longest red waves can penetrate Venus' dense atmosphere. Arc at bottom of picture is base of lander, white arm and twin tubes at right above.

km) apart. Space stations hovered above the planet and relayed data back to Earth.

The Venera 12 lander sent back data for a record 110 minutes. It reported an air pressure eighty-eight times Earth's and a temperature of 860°F (460°C). Venera 11 broadcast data for ninety-five minutes. Both spacecraft saw lightning flashes in the atmosphere.

A lander's-eye view must be weird indeed. Only about 10 percent of the sunlight striking Venus reaches the ground. The sun would not be visible in the sky. Only the longest red light waves in sunlight can travel any distance through the dense atmosphere, so illumination is red. Visibility in this dense carbon dioxide air would be only about one and a half miles, and everything would look distorted due to refraction.

A robot explorer called the Venus Orbiting Imaging Radar (VOIR) is on the NASA drawing board for launch by a space shuttle in 1984. VOIR would subject Venus to the most extensive scientific examination ever while circling the planet for at least seven months.

Optimistic futurists foresee human colonies on Venus in a few hundred years. The planet would have to be "terraformed"—made more Earth-like—first. Humans could seed Venus with simple life forms, anticipating nature by billions of years. Extremely hardy algae injected into the atmosphere might convert carbon dioxide into organic sub-

stances and oxygen by photosynthesis. Water might be condensed out of the atmosphere or released from rocks, and raw materials mined. In view of the stunning technological feats of the twentieth century, who would dare to say that such things will not be possible?

Mercury

Mercury is an unlikely world to harbor living creatures. At thirty-six million miles (58 million km), it is the closet planet to the sun. Mercury zips around the sun in eighty-eight days, but takes fifty-eight and two-thirds days to spin once around its axis. This places first one hemisphere and then the other toward the sun at perihelion. With a negligible atmosphere, Mercury is alternately torrid and frigid.

Observing Mercury from Earth is difficult be-cause the planet is only 3,000 miles (4,800 km) in diameter and always near the bright sun. Our best look at the planet so far has been through the robot eyes of the Mariner 10 spacecraft. The 1,042-pound (473-kg) Mariner 10 first encountered Mercury on March 29, 1974. It flew within 450 miles (720 km) of the unlit hemisphere carrying a twin-camera TV imaging system and instruments to measure solar wind, temperature variations, and surface composition.

Afterward, Mariner 10 and Mercury kept on circling the sun in their predetermined orbits.

Mariner 10 was maneuvered into a second encounter with Mercury on September 21. Then it flew 30,000 miles (48,000 km) above the planet and photographed different areas. A third encounter on March 16, 1975, took Mariner 10 within 204 miles

Above: Pioneer 2 transporter glows as it enters Venusian upper atmosphere. Four other probes it carried have been ejected and will transmit data until destroyed. One tough capsule (opposite bottom) operated for 67 minutes it survived in heat of Venus' surface. Opposite top: Radar mapping by Venus Orbiter disclosed largest canyon yet found in solar system: some 5 km deep and 1,450 km long. These pictures are artist's conceptions. In carbon-dioxide air of Venus, visibility would be lower and images distorted.

(328 km) of the surface. Then its small control jets ran out of gas, so Mariner 10 will send no more data, even though it still passes Mercury periodically.

Mariner 10 returned almost three thousand pictures of Mercury. Craters ranging up to several hundred miles in diameter look as if they had been blasted out by meteorites. The largest is Caloris Basin, which is about 930 miles (1,500 km) in diameter. It has unusual fractures and ridges, and is ringed by cliffs that look like bedrock uplifted by a cataclysmic impact.

Some craters are filled with smooth lava. The youngest craters have bright rays radiating outward. There are smooth plains that look younger than the heavily cratered areas yet older than the rayed craters. These are crossed by many ridges and long lines of cliffs. The mountains may have resulted from lava flows of rocks melted by impact energy or from volcanic activity. An infrared radiometer found a few inches of dust covering the surface.

A magnetometer found a very weak magnetic field that deflects the solar wind around Mercury so high-speed particles do not bombard the surface. Mercury's magnetic field apparently originates within the planet. It is a one-thirtieth-scale replica of Earth's magnetic field.

Mercury is rugged and inhospitable, but not impossible for future explorers. It would make an excellent solar research station. Temperatures range from 800°F (427°C) at local noon to −300°F (−184°C) at midnight. Ultraviolet spectrometer observations have revealed a tenuous atmosphere—with traces of helium, argon, oxygen, carbon, and xenon—that

Opposite: Inert and inhospitable Mercury was closely observed by Mariner 10 (above). As seen from 500,000 (top) and 88,000 km, surface resembles our moon's. It is pocked with craters, the largest some 1,500 km across. Overlapping craters (bottom) are 40 and 80 km/D. Atmosphere is negligible, so temperatures are extreme: 427°C at local noon, −184°C at midnight. Above: Mariner 10 has standard spacecraft elements: solar panels for energy, dish antenna for communication with Earth, various booms for instruments and radio antennas.

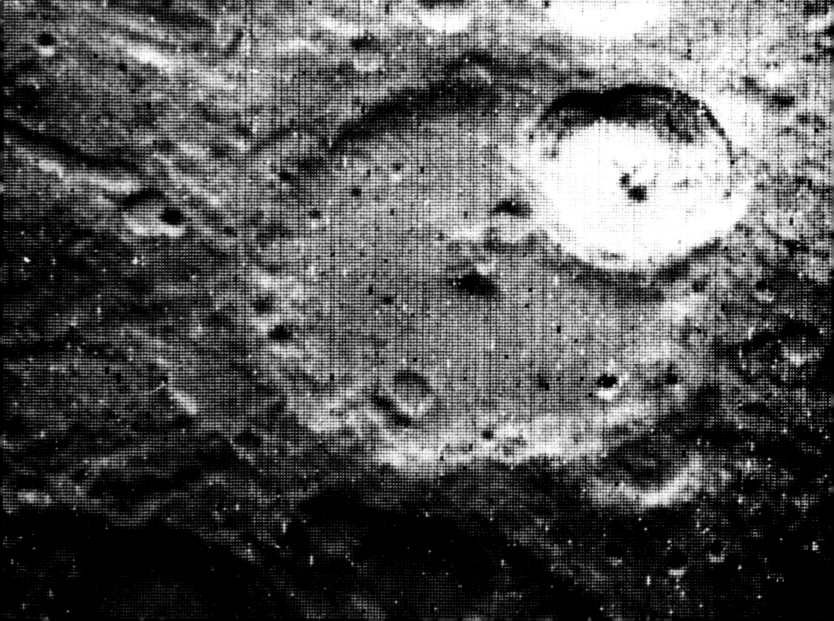

Brilliant and exciting pictures of giant Jupiter and several of its many moons were taken by Pioneer 10 and 11. Top: From 20 million km, Jupiter with Io (left) and Europa. Io is some 350,000 km above planet's Great Red Spot, possibly a cyclonic storm center. Europa is 600,000 km above a Jovian cloud mass. Far left: Closeup (from 2 million km) of Europa shows unusual crisscrossed linear structures, possibly crustal fractures. Left: Ganymede from 2.6 million km. Larger than our moon, it is probably composed of rock and ice.

is a trillion times thinner than Earth's. We can expect the first scouts to be robots, perhaps followed a few hundred years later by humans.

Asteroids

Thousands of rock chunks called asteroids or minor planets have been orbiting our sun since its birth. Asteroids are unadulterated samples of the original matter that formed our solar system. Some asteroids contain water and carbon. These provide strong evidence that the raw materials of life have existed beyond Earth for billions of years!

Most of the asteroids orbit the sun inside a region called the asteroid belt that stretches between the orbits of Mars and Jupiter. More than 2,000 asteroids of various sizes have been catalogued and millions of tinier ones probably exist. The largest of all is Ceres, which is 480 miles (770 km) wide. The total mass of all the asteroids—from dust grains to those several miles wide—is probably much less than our moon's.

Metals and water locked in asteroids could be a tremendous resource for space colonists, or even for the depleted Earth of the future. NASA scientists are developing an ion-drive propulsion system that uses solar energy to operate at low thrust for many years. It is expected to be ready for visits to the asteroids in the 1980's.

Over twenty-five asteroids called Apollos are known to fly orbits that come inside the Earth's orbit. They are the most logical targets for early visits, since some cruise within a million miles of Earth. An unmanned robot spacecraft could rendezvous with Ra-Shalom, which was discovered in 1978. This roughly two-mile-wide (3-km) rock chunk orbits the sun every nine months, traveling from outside Earth's orbit to within Venus'. It looks like a carbonaceous chondrite type that is rich in organic materials.

Robot spacecraft could photograph asteroids closely and prospect for resources as early as the 1980's. By the 1990's, they should be able to make round-trip flights. They would land, collect, and bring back rock samples. In the twenty-first century astronauts might attach spaceships like towboats to asteroids to push them into orbits as needed for intensive mining operations.

Jupiter

Giant Jupiter, or some of Jupiter's moons, may have the critical compounds of life or even microbes. Jupiter's environment resembles the primitive one in which life theoretically developed on Earth. Energy is plentiful and there has been ample time for complex compounds to form.

Jupiter is the fifth planet from the sun. At a distance of 483 million miles (777 million km), it orbits the sun once every twelve years. Jupiter spins around its axis in nine hours fifty minutes, which causes the planet to bulge at its equator and be flattened at its poles. Jupiter is more massive than all of the other planets combined. Compared to Earth, it is 318 times more massive and eleven times wider. It could hold more than thirteen hundred Earths inside. Jupiter's strong gravity holds on to thirteen or more moons, each of which orbit

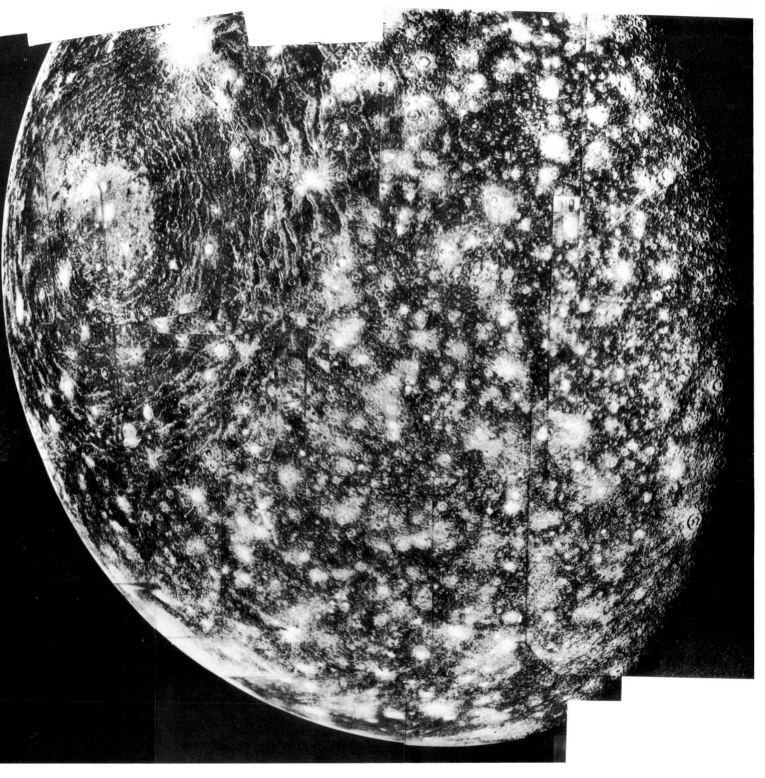

Callisto, darkest and most distant of Jupiter's Galilean satellites, as seen by Voyager 1 from 202,000 km. Photomosaic shows surface heavily cratered by meteorite impacts, suggesting that Callisto may have oldest surface of any Galilean moon. Large bright spot (upper left) is an impact basin some 600 km in diameter. Concentric rings extending more than 1,000 km probably resulted from shock waves produced by mighty impact on icy surface. Of greatest interest has been variations in condition and composition of moons.

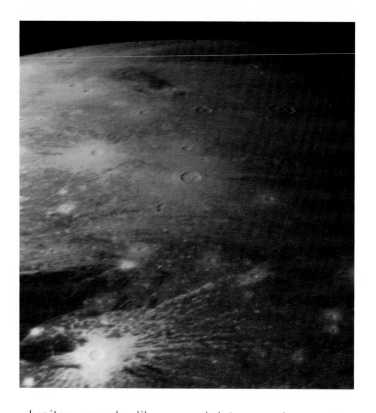

Jupiter, much like a miniature solar system.

Close-ups of Jupiter were first snapped by the robot spacecraft Pioneer 10 and 11 as they sped by the giant world in 1973–74. These robot adventurers will be the first earthlings ever to go outside our solar system. Pioneer 10 will leave first in 1987. Just in case they encounter intelligent space creatures, both spacecraft sport gold anodized-aluminum plaques that picture when, where, and by whom they were launched.

In the late 1970's Jupiter, Saturn, Uranus, Neptune, and Pluto were all lined up in a row in space, a phenomenon that occurs only once every 171 years. They were last so aligned when Thomas Jefferson was President and will not be again until 2147 A.D. The Voyager 1 and 2 robot probes were launched in 1977 to take advantage of the outer planets' unusual alignment. First the Voyagers encountered Jupiter in 1979. Jupiter's gravity has slung them on to Saturn for encounters in 1980–81. Saturn's gravity may be used to hurl them on to Uranus for fly-bys in 1986. Finally Uranus' gravity can assist for a Neptune fly-by in 1989. The NASA travel budget was not adequate to include outermost Pluto on this itinerary.

At the time of their launching, Voyager 1 and 2 were the most sophisticated robot probes ever undertaken. They were working farther from Earth than had robots ever before. They were more independent of Earth-control than any of their predecessors. The great distances (billions of miles) and long time (more than eight years) of the journey have required that the Voyagers be able to care for themselves and perform complex scientific surveys with-

Opposite: Great Jupiter—more massive than all our planets combined—from 4.3 million km. At top is Red Spot, here three times the size of Earth, and white ovals, first seen to form in 1939–40. Swirls are atmospheric detail. Far right: Region east of Red Spot from 1.8 million km. Differences in cloud color may indicate relative height of layers. Right: Same area in exaggerated colors produced by computer to enhance variations. Above: Ganymede (left) shows smoother surface than Callisto. Io (right) has volcanic eruption occurring on rim.

out continual and specific directives from home.

The Voyagers each weigh 1,820 pounds (825 kg). They have camera systems to take hundreds of color pictures; radio astronomy experiments to examine energetic radio outbursts; and ultraviolet, infrared, and charged-particle sensors. Their one-foot-diameter (30-cm) antennas for communicating with Earth are the largest yet flown on planetary missions. Since the Voyagers fly too far from the sun to depend on solar energy, nuclear power is provided by radioisotope (plutonium 238) thermoelectric generators.

The Voyager slow-scan vidicon cameras convert light to electrons that are scanned and processed into digital form for transmission to Earth. Each Voyager picture is a composite of 800 lines with 800 elements per line. The Voyager uses a very high frequency signal to radio home 115,200 bits per second, or a complete picture in forty-eight seconds.

Voyager 1 flew within 174,000 miles (280,000 km) of Jupiter on March 5, 1979, and sent back spectacular views of Jupiter and some of its moons. The spacecraft discovered a thin ring of rocky debris encircling Jupiter 34,000 miles (54,400 km) out. The ring looks more than 5,000 miles (8,000 km) wide and somewhat less than eighteen miles (29 km) thick. It had never been seen from Earth because it is thin and transparent, except when sighted straight on.

Thick stripes of swirling multicolored clouds blanket the huge planet. Great storms and convoluted weather patterns circulate in the atmosphere.

Jupiter radiates twice as much heat as it gets from the sun. Internally generated heat probably drives the turbulent gases. The famous Great Red Spot is Jupiter's highest atmospheric feature. It looks like a violent cyclonic storm center. Changes in the maelstrom were recorded for several days. The churning, contorted spot is actually so big that several Earths could easily be held within it.

Apparently Jupiter has no solid surface. It is a rapidly spinning ball of gas and liquid (mainly hydrogen) with perhaps a small molten iron-silicate core. It has a very favorable environment for life to evolve. The thick atmosphere is mostly hydrogen and helium, with traces of methane, ammonia, water vapor, and other gases. An intense aurora at least 18,000 miles (28,900 km) long with superbolts of lightning was discovered by Voyager 1. Lightning in the Jovian atmosphere could initiate the formation of complex biological molecular units.

Four of Jupiter's moons, discovered by Galileo Galilei in 1610 and collectively named the Galilean moons, first became meaningful worlds in Voyager 1 photographs. The inner two, Io and Europa, are about the size of Earth's moon. Ganymede and Callisto are even larger than the planet Mercury. Little red Amalthea, Jupiter's closest moon, remains an enigmatic red blob.

Brilliant red-and-gold Io looks startlingly smooth, without impact craters but with the first active volcanoes ever seen outside Earth. Most spectacularly, eruptions were photographed sending greenish gas plumes and other material more than 150 miles (241 km) above the surface. Intensive

Further exploration of Jupiter could be undertaken after 1,000-day flight by orbiter-probe spacecraft. Painting shows probe unit being launched, with 56 flying days to go, so its trajectory will take it into planet's atmosphere on sunlit side. During 30-minute descent to surface, probe will measure atmosphere. Orbiter would be expected to circle Jupiter for at least 20 months to conduct thorough study of planet itself, its largest satellites, and entire environment, including surprising ring of rocky debris.

Opposite: Schematic drawing shows Voyager circling Saturn two years after departing Jupiter. Though famous for its rings and most distant planet known before the telescope, Saturn is 1,426 million km from Sun and perceived in observatory photographs, like that at bottom left. Above: View of Saturn from Titan, largest of its ten moons and the most interesting to exobiologists. Titan is near Mercury in size, has substantial atmosphere, perhaps as dense as Earth's, and a solid surface—all suggestive of life.

study of photographs also revealed "blue snow"—probably wisps of gas venting from fractures in Io's crust and condensing into crystals.

Orange-tinted Europa's most distinctive feature is a crisscross of linear structures more than 100 miles (161 km) wide and 1000 miles (1609 km) long. Probably they are geological faults.

Ganymede, Jupiter's largest moon, is believed to be covered by water-ice mixed with rock and may be as much as 50 percent water by weight. It has oddly sinuous ridges and grooves in its surface which may be evidence of internal heavings, like Earth's. It is heavily marked with impact craters, like our Moon, but has no mountains or "major relief."

Callisto, more than a million miles out from Jupiter, has an enormous, multiringed meteor impact basin in its pink-and-brown icy face, suggesting different crustal properties.

Saturn and Beyond

In Earth-bound telescopes, Saturn resembles Jupiter. It has alternate dark and light cloud belts with transitory spots resembling hurricanes. Hydrogen and helium gases predominate in the atmosphere, with traces of methane, ethane, and ammonia. Saturn radiates nearly twice as much energy as it gets from the sun. Thus it is not unreasonable to speculate that organic molecules or even primitive life forms have evolved there.

Four dazzling rings circle Saturn. Probably composed of a swarm of jagged rocks orbiting the planet, they shine by reflecting sunlight. Saturn was the most distant planet known before the telescope. Its orbit is some 886 million miles (1,426 million km) away from the sun, where it receives only one one-hundredth the sunshine that Earth does.

Saturn is 75,000 miles (120,000 km) in diameter. It has a volume 815 times greater than Earth's but only ninety-five times the mass. That gives Saturn the lowest density of all the planets. It could float in water if there were a large enough sea somewhere. Saturn also has the greatest polar flattening of any planet, probably accentuated by its rapid ten-hour, fourteen-minute rotation period.

With its ten known moons, Saturn orbits the sun every 29½ years. Titan is the largest of the moons and the most interesting to exobiologists. With a diameter of 3,600 miles (5,760 km), Titan is comparable to the planet Mercury. It has a substantial atmosphere that may be as dense as Earth's. The gases methane, ethane, and acetylene identified in its atmosphere, plus a solid surface, make Titan seem a likely site for life to start. If complex organic compounds formed in the atmosphere, they could drop to the surface and accumulate. Primitive organisms could be thriving there today!

If NASA exercises the option to fly Voyager 2 on to Uranus, the spacecraft will encounter the seventh planet from the sun in January, 1986. Uranus is 30,000 miles (48,000 km) in diameter. Because it is 1,780 million miles (2,850 million km) from the sun, Uranus is still very enigmatic. In contrast to the other gas giants, Uranus is tilted so that it lies on its side as it orbits the sun every eighty-four years. This extraordinary position puts the north

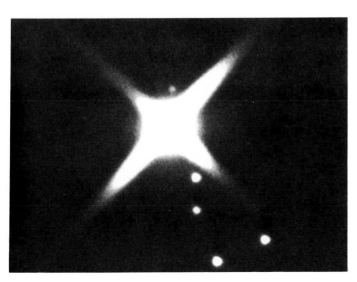

pole directly in sunlight every eighty-four years and in the dark forty-two years later as the sun shines on the south pole. In 1986 the orientation will permit Voyager 2 to fly almost vertically through the equatorial plane, where the satellites of Uranus are.

Voyager 2 could look closely at the rings of Uranus, which were discovered in 1977, and photograph the planet's sunlit side, plus several satellites including the largest, Miranda. It could sight any magnetic field and plasma clouds that may be present. After encounter, Voyager 2 could fly out through the planet's shadow and look back at the dark side.

Finally, Voyager 2 may look at Neptune in 1989. At 2,790 million miles (4,460 million km) from the sun, Neptune is a virtually unknown world. So far, hydrogen and methane have been detected in its atmosphere. Neptune is 28,000 miles (45,000 km) in diameter with two known moons, Triton and Nereid. It orbits the sun every 165 years and spins on its axis every fifteen hours.

After completing their missions, the Voyagers will cross the orbit of Pluto in 1989. As they were not funded to probe the sun's outermost planet, such basic data as Pluto's size, mass, and whether it is actually a planet at all, will remain uncertain. Pluto's orbit crossed inside Neptune's in 1978 and for the rest of this century Pluto will be closer to the sun than Neptune. An apparent moon was discovered in 1978, but little is known about these two small, distant worlds. They are probably frozen and lifeless, appropriately named for the god who ruled the mythological world of the dead.

The Voyagers will escape the solar system at a speed of 38,700 miles (62,000 km) per hour. Even at that rate they will not approach another star for at least another 40,000 years in the vast emptiness beyond Pluto. Other predictable Voyager approaches to stars will occur in 147,000 and 525,000 years. On the chance that intelligent spacefarers might encounter the Voyagers, astronomer Carl Sagan of Cornell University put a phonograph record, cartridge, needle, and playing instructions aboard both of them.

The twelve-inch (30-cm) copper disk is encased in an aluminum cover so it can survive more than a billion years in empty space. It records the best of planet Earth and its people. There are, in scientific language, pictures of the Earth and our location in space, human beings in various settings, and the Voyager mission. Spoken greetings in about sixty languages and a sound essay including such sounds as weather and surf, birds, and human civilization follow. Finally, there are ninety minutes of musical selections representative of the marvelous cultural diversity of Earth.

In the next millennium it should be possible to send robot spacecraft to look for extraterrestrial life near likely stars. The experience gained in the exploration of our own solar system will enable scientists to build more sophisticated probes for interstellar space. The robots could go into orbit around possible inhabited planets and even drop automated landers. Scientific studies like those carried out by robot scouts here will pave the way for human exploration of other likely planetary systems.

Uranus is enigmatic but may yield secrets if Voyager 2 arrives in 1986. At far left is observatory photograph overexposed to show five satellites. At left is drawing of planet's recently discovered rings. In deep beyond lie Neptune, virtually an unknown world, which Voyager 2 may approach in 1989, and outermost Pluto, so remote it is not even certainly a planet. After that, Voyagers will escape solar system and head for stars. Though traveling at 62,000 km/hr, closest star will be 40,000 years distant in space.

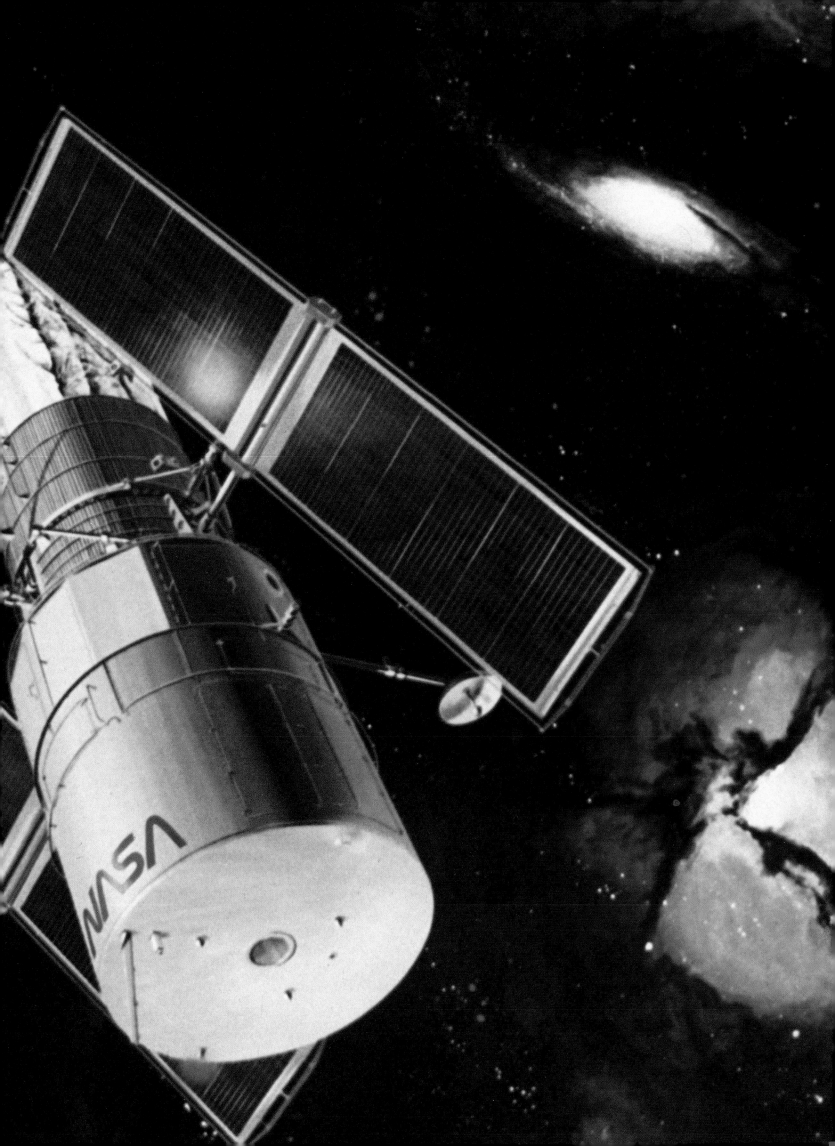

Looking for Other Planetary Systems

All information about the populace of deep space is elicited from its radiations. The human eye sees only a fragment of all the radiant energy that bombards Earth. Modern technology extends our vision to a whole spectrum of invisible rays that could one day reveal distant inhabited worlds. Today in addition to light, a wide range of radio, infrared, ultraviolet, X, and gamma rays are being tapped for clues.

Although planets theoretically originate in the same contracting nebulas as do stars, they accrete much less body material. They do not have sufficient mass to fire their own internal nuclear reactions. Planets are illuminated by a nearby star. Any planet alone in space without a star would not shine at all. It would be undetectable except for an accidental collision with a spacecraft.

The first step in detecting other planetary systems, regardless of which rays are being tapped, involves collecting sufficient radiation for analysis. Data are commonly processed by preprogrammed electronic computers that can perform calculations with superhuman speed. Computers may display processed data on video screens, store results in their memory banks, or enhance input and generate pictures drawn from invisible radiation.

Optical Telescopes

It is possible that powerful optical telescopes will soon be able to glimpse planets circling neighbor suns, although they cannot do so yet. Our blazing sun outshines Jupiter by a factor of 250. Even the world's best telescopes cannot pick out a planet in such a glare at the vast distances involved. However, they do provide persuasive indirect evidence that such planets exist.

Optical observatories are busier than ever as astronomers look for fainter and fainter objects in space. Giant reflecting telescopes have parabolic mirrors to collect and focus feeble starlight on photographic plates or electronic recording devices. There it is accumulated until it is bright enough for analysis. The mirror is made of materials such as fused quartz and ceramics, so its surface won't be distorted by temperature fluctuations. Its reflecting surface is commonly a thin layer of aluminum.

The largest telescope in the world is the 236-inch (6 m) reflector at the Special Astrophysical Observatory in the U.S.S.R. (Telescope size refers to mirror diameter.) The largest in the United States is the 200-inch (5 m) Hale telescope on Mount Palomar in California. These giants can picture stars that are a million times fainter than the human eye can perceive. Using time exposure, they can photograph stars as faint as a candle viewed from 10,000 miles (16,000 km) away.

A clear dark sky is best for sky searches. All major observatories are located high up in mountainous regions where the air is relatively thin, dry, and free from dust and light pollution. The 158-inch (4-m) Mayall telescope was opened in 1973 at the Kitt Peak National Observatory, about fifty miles from Tucson, Arizona, following an extensive hunt for a prime location in the United States. This telescope has recently provided photographic evidence of possible planetary systems in deep space.

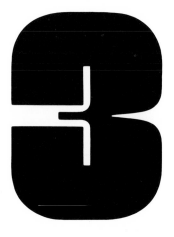

Preceding pages: Artist's concept of NASA Space Telescope in orbit at 500-km altitude, free of Earth's obscuring atmosphere. Two solar panels provide power for on-board computers and miniaturized instruments. Above: Astronauts make adjustments as space shuttle prepares to launch 11-t, 13-m observatory. It will be operated from Earth by remote control and probe new depths of space, making first-step observations to locate life-bearing planets elsewhere.

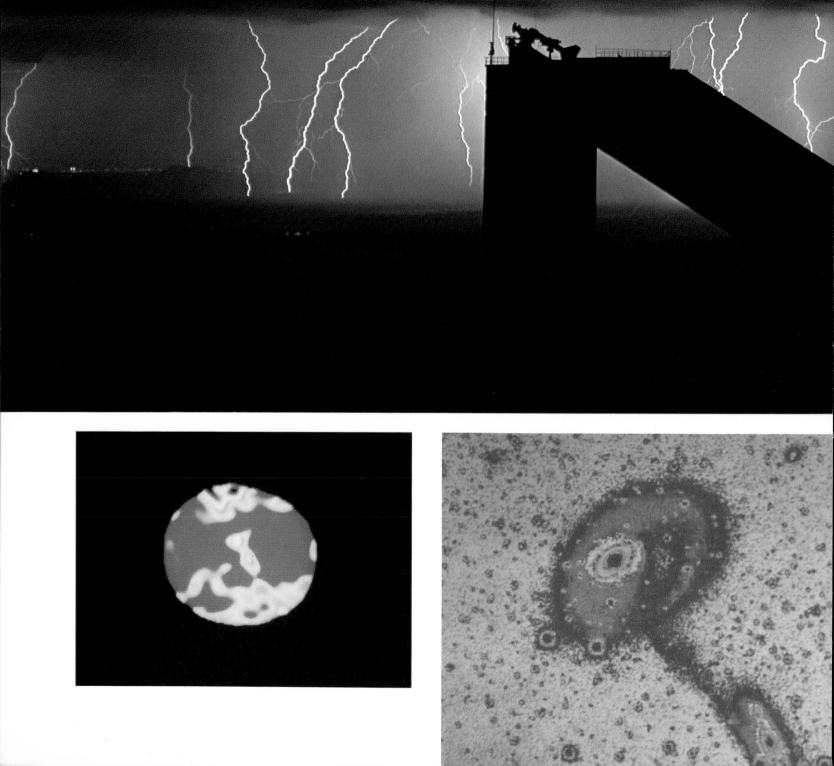

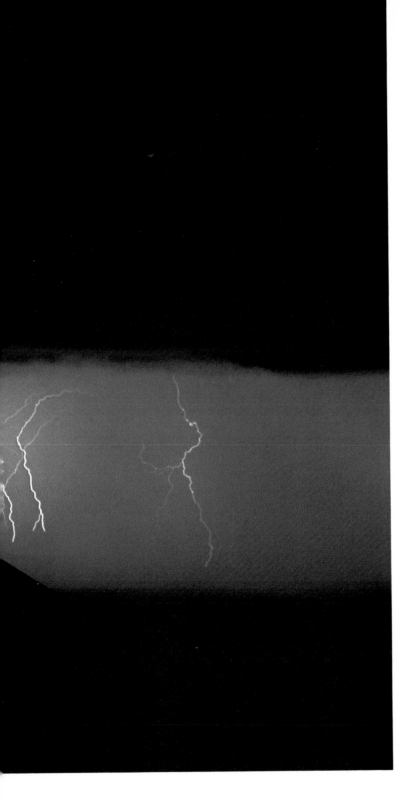

Much recent construction of optical observatories has occurred in the southern hemisphere because many interesting sky objects are only (or best) visible there. The 158-inch (4-m) reflector at the Cerro Tololo Inter-American Observatory in the Andes Mountains in Chile is the largest telescope located below the equator. Next in size is the 153-inch (3.9 m) Anglo-Australian reflector at Siding Spring, Australia.

Telescopes need greater light-gathering power than the biggest mirrors can provide to see planets near dazzling stars. The cost of still larger single mirrors is prohibitive, so alternative technology is being utilized to discern ever fainter celestial objects.

At the most advanced observatories, light collected by the telescope is focused onto electronic image intensifiers. These generate an electric current that is amplified. The information can be recorded on photographic film, displayed for viewing on a video screen, or monitored and processed by a computer. Light-sensitive solid chips such as those commonly used in hand calculators can be put in an array called a pixel to register very faint light and to transmit it to a processing computer.

The Kitt Peak National Observatory turns digital arrays of data into spectacular pictures. The pictures are either representative of the actual appearance of the light source or a color display emphasizing particular features. The Observatory's interactive picture processing system (IPPS) uses a television screen and an electronic pointer so astronomers can easily select for later analysis the

Splendid lightning discharge illuminates solar telescope of Kitt Peak National Observatory complex near Tucson, AZ. Astronomers accumulate feeble photons—light-energy units—from objects many light-years away until they are bright enough for analysis. Electronically intensified images produce arrays like those of red star Betelgeuse and blue pair of interacting galaxies. Major problem in detecting planets is glare from star—or sun—providing its illumination. Telescopes with greater light-gathering power may overcome this.

pictures that look as though they may be important.

Possibly the forerunner of future mighty telescopes, the Multiple Mirror Telescope (MMT) synchronizes light collection by six seventy-two-inch (183-cm) mirrors on a common mount. The six perform together as a single powerful 176-inch (447-cm) mirror on a mountaintop near Kitt Peak. Because all six mirrors must direct their light exactly to a perfect single focus, the MMT was built with enormous precision and extraordinary support and housing. The housing weighs 500 tons, (453 t), yet it rotates smoothly when the six reflectors move together. The MMT was developed by the Harvard-Smithsonian Center for Astrophysics and the University of Arizona.

A 400-inch (10 m) telescope is on the drawing board at the University of California at Berkeley. Two different designs are being explored. The first would follow the MMT lead by linking several smaller mirrors together precisely. The second would pioneer a plastic mirror whose curved surface could be maintained free of distortion by a computer-regulated movable support.

Indirect Detection Methods

To date astronomers have relied on observations of visible stars to infer the presence of gravitationally bound planets. Instruments reveal that stars are racing about in space. Stellar motions are analyzed in two parts: movement across the sky, called proper motion, and approach or recession, called radial velocity.

Astrometry is the precise measurement of the positions and motions of visible stars. It is an old technique currently in favor and in use because of the increasing refinement and sophistication of the instrumentation. The measurements require many years of painstaking effort to reveal unseen planets. Sir Isaac Newton's laws of motion predict that a lone star will travel a straight path across our sky, while a star with companions will wobble due to their gravitational tug.

There are about fifteen rather small nearby stars whose motions across the sky can currently be determined precisely enough to reveal a massive planet like Jupiter. Measurements are extremely difficult, even for close stars. Stellar motions are barely perceptible across the immense distances separating us. Measuring the typical wobble is like detecting a one-inch movement at a distance of 100 miles (160 km)! Previously unseen dwarf stars in binary systems have been found by astrometry. In 1844 this method revealed the white dwarf Sirius B accompanying brilliant Sirius A through space. Planets are still controversial.

Telescopes used to study proper motion are different from the giants used to look at really faint objects. A field of view large enough to include comparison stars, photographic records with measurable displacements, great stability, and a system that is not changed over a period of many years are critical tools for astrometric studies.

The Sproul Observatory in Swarthmore, Pennsylvania, has for many years been a research center for detecting unseen companions of nearby stars. Starting in 1937, astronomer Peter van de

Three observations of X-ray sources by Einstein (HEAO 2) satellite: O stars (top) in gaseous nebula Eta Carinae, Crab nebula (bottom left), and Quasar 3C273 (right). O stars are young, the Crab nebula is remains of supernova. Quasars, strange objects that radiate huge amounts of energy for their size, appear to be billions of light-years from Earth. High Energy Astronomy Observatories are penetrating deep space to identify and analyze X-ray emissions from these sources as clues to how the cosmos evolved and life processes began.

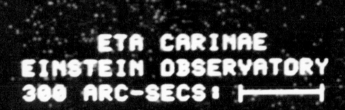

ETA CARINAE
EINSTEIN OBSERVATORY
300 ARC-SECS: ⊢———⊣

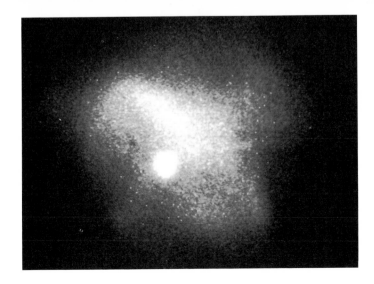

Kamp photographed Barnard's star with the Sproul's twenty-four-inch (61-cm) refracting telescope on 766 nights during the next eighteen years. Barnard's star is a faint red dwarf star located some six light-years away in the constellation Ophiuchus. Only 1/2,300 as luminous as the sun, it is the sun's second closest neighbor star. It is an especially good target for a planet search because it moves across the sky faster than any other known star.

Even for Barnard's star, changes in position don't become apparent for years. Although it is zooming along at 198,000 miles (317,000 km) per hour, we would need 170 years to notice a change in its position equal to the moon's diameter. By the end of 1968, Van de Kamp had examined 3,036 photographic plates containing 10,452 exposures of the progress of Barnard's star.

He has announced a minuscule wobble in the motion of Barnard's star that looks as if it is caused by the gravitational force of two massive planets like Jupiter. (His study was not sensitive enough to reveal a planet like Earth, which is only 1/318 as massive as Jupiter and would not tug as hard.) Recent refinements of astrometric techniques have raised doubts about the wobble. At present astronomers are uncertain whether or not planets circle Barnard's star.

The astrometric search for planets continues. In 1977 astronomer Sarah Lee Lippincott announced the results of applying Sproul's new automatic measuring techniques to forty years of observations of a faint red dwarf star designated BD+68°946. She found a tiny wobble, again suggesting the presence of a planet.

Careful analysis of starlight adds further indirect evidence that other planetary systems exist. Starlight is composed of a mixture of different-length waves or colors. A device called a spectrograph is often attached to the large reflecting telescopes to split starlight into its spectrum of component colors. If a visible star has unseen planets attracting it gravitationally, changes in its radial velocity will theoretically show up in its spectrum within three years.

Radial velocity can be determined from a star's spectrum because of the Doppler effect, named for its nineteenth-century discoverer, physicist Christian Doppler. Spectra of receding stars contain longer waves (red shift) than spectra of identical glowing gases in the laboratory. Conversely, the spectra of approaching stars have

World's largest optical telescope is 6-m reflector at the astrophysical observatory of USSR Academy of Sciences. Located on a 2,170-m peak near Zelenchukskaya in the Karachaevo-Cherkesskaya Autonomous Region of the Caucasus, telescope is capable of photographing stars a million times fainter than human eye can see, or as faint as a candle viewed from 16,000 km. Sites are chosen for thin, dry atmosphere free of dust, extraneous light, vibration. Scale of installation can be judged by size of man leaning over walkway (right).

shorter waves (blue shift). Recent spectrographic surveys by astronomers Helmut A. Abt and Saul G. Levy at the Kitt Peak National Observatory found changes in radial velocities for one-tenth of their sample of 123 stars like our sun. These changes may be caused by the gravitational tug of planets. Further, seven out of forty-two massive hot stars in a subsequent study showed orbital motions indicative of planets. The results could mean that up to 20 percent of the billions of multiple star systems in our galaxy include planets!

Infrared Astronomy

Infrared telescopes may get the first look at planets circling other suns. They frequently scrutinize radiations from planetary atmospheres, newly forming stars with possible planets, and the planet-breeding interstellar clouds.

Theoretically, planets are much more distinguishable from their star in infrared light than in visible light. Planets absorb starlight, warm up, and shine back infrared rays. A planet's infrared brilliance is proportional to its surface area and temperature. Jupiter looks 100 times brighter in infrared than in visible light when compared with the sun.

Dust grains in contracting clouds also transform starlight to infrared rays. They shine over a much larger area than do resulting solid planets. Observations of the minerals of rocky planets in orbital motion would be strong evidence of new worlds to come.

Dying suns throw off dust and gas enriched with chemical elements to be recycled into new stars, planets, and creatures. Gossamer rings encircling old stars glow at infrared wavelengths. They show which materials shape the next generation.

Unknown worlds could be discovered at any time of the day or night. Sunshine does not blot out infrared stars, so the telescopes work a full twenty-four hours daily. Observing sessions are highly automated. The human eye cannot see infrared sources. Instead of sighting stars, an astronomer selects the source to be observed and commands a computer to get data. Computers point and control the telescopes, apply corrections for tracking, and collect data which they store, print out, or visually display for analysis by astronomers.

Water vapor in the air strongly absorbs infrared rays. Consequently, observatories are located on very high mountain peaks selected for the dry clear sky overhead. The world's largest infrared telescope is the United Kingdom's 3.8-meter (150-inch) reflector on the 13,800-foot (4,140-m) summit of the extinct volcano Mauna Kea in Hawaii. Nearby is the National Aeronautics and Space Administration's three-meter (119-inch, 302-cm) Infrared Telescope Facility. Oxygen is so rare at that altitude that visiting astronomers could suffer impaired mental and physical efficiency. A special dormitory has been built at the 9,000-foot (2,700-m) level to provide altitude acclimation.

Infrared telescopes resemble other large reflectors in optical configuration. However, they require a special heat sensor called a bolometer because photographic plates register only visible light and shorter wave radiation. Bolometers made of

Largest American reflector is 5-m Hale telescope on Mount Palomar in California. Top: Huge structure points north. Bottom: Domed housing. A 10-m telescope is projected but outsize single-unit quartz-and-ceramic reflectors are prohibitively expensive and extremely difficult to make. New giant could link smaller mirrors, like MMT, or pioneer use of plastic. Search for planets is pursued by astrometry, which is measurement of visible stars' proper motion and spectroscopy, radial velocity of approach or recession from Earth.

germanium doped with gallium are commonly arrayed to detect different ranges of infrared wavelengths. They scan a source transversely and gather data for broad, informative pictures. The biggest infrared telescopes so equipped could now reveal a lone planet the size of Jupiter at a temperature of 81°F (27°C) if it were situated like the close stars.

Everything—telescopes, human bodies, equipment, and observatory walls—gives off infrared rays. If this unwanted heat gets to the bolometers, they go haywire. The Hawaii infrared telescopes and the MMT have baffles to protect a sensor at the focus from stray heat. Both the baffles and detector are cooled with liquid helium to a level a mere four degrees above absolute zero.

In addition, a beam switching technique is used to ensure that stray heat in the sky does not obliterate faint infrared sources. Two small neighboring patches of the sky are repeatedly observed in rapid succession. The first patch includes the source and the second does not. The resulting electronic signal distinguishes between source and background radiation.

Small infrared telescopes are lofted above the atmosphere's obscuring water vapor in planes, balloons, and rockets. They find sites that are undetectable by larger telescopes at lower altitudes. A one-inch (2.54-cm) balloon-borne infrared telescope first discovered immense cool dust collections at our galaxy's center. Future observations from spacecraft such as the projected Infrared Astronomical Satellite Survey should uncover many new dust clouds, stars, and possibly planets.

Six Shooter: Innovative MMT is world's third largest telescope, but lighter, easier to handle, and less costly than older designs. Since focal length is essentially that of 1.8-m reflector, MMT is also extremely compact. Telescope and four-story housing are a single 550-ton unit which rides a circular track on four steel wheels. Laser guided sensing system automatically controls tilt and position of mirrors, corrects focusing errors. Near Kitt Peak, MMT excels in infrared photometry and other nonvisible-light observations.

Space Telescope

The best telescopes on Earth are blocked by an obscuring veil of air from seeing even a Jupiter-sized planet circling a nearby star. Airless outer space beckons planet seekers!

Space Telescope promises the most exciting views ever. This ten-ton (9-t) observatory is expected to orbit a ninety-four-inch (239-cm) telescope 310 miles (500 km) above the ground by 1983. Hidden worlds could easily come into view as it probes 350 times deeper into the darkness than is now possible.

Space Telescope is expected to serve as an observatory in space at least into the 1990's. All of the instruments on board will be miniaturized and lightweight yet strong enough to survive the shocks of launch. Electrical power will be obtained from sunlight using solar arrays. Astronomers will operate the observatory by remote control from Earth. Astronauts will maintain and change instruments in orbit. They will retrieve it for return to Earth by space shuttle when extensive overhaul is necessary.

Future space observations with improved detection techniques promise tantalizing surprises. Certainly improved electronic, astrometric, and spectrographic systems in space will have a superb chance of revealing distant planets indirectly.

High Energy Astronomy

High energy, short wave, ultraviolet, X, and gamma rays are strongly absorbed by the atmosphere. Since these rays could be lethal, we are lucky to be shielded from their danger. But it is no wonder high-energy rays entice space-age astronomers. They alone bear clues about the hottest stars, the strangest cadavers, and the most powerful energy sources in the universe.

X-rays have thousands of times the energy of ordinary light. Gamma rays have millions of times more energy. On Earth these rays are produced by naturally radioactive minerals, as well as by technology in known physical processes. In deep space extraordinarily energetic X-rays and gamma rays are generated in processes not yet comprehended. Fiery X-ray stars are frequently thousands of times brighter than our sun.

Pulsars are stars that emit radio signals in extremely precise pulses. They seem to be fast-spinning compact balls of neutrons left over from a great supernova. A black hole is theoretically the last stage in the collapse of a dying massive star. It is so densely packed that its gravitational force is great enough to prevent everything—even light waves—from escaping its confines.

Quasars are estimated to pour out 100 times more energy than a typical galaxy of 100 billion stars, yet be only 1/10,000 as big. Quasars are the most powerful energy emitters known. How this enormous energy is generated is a mystery. Pulsars, black holes, and quasars are all at the frontier of high-energy astronomy.

High-energy telescopes focus extreme ultraviolet and some X-rays by reflection at grazing incidence from a series of concentric metal collars around the beam. They commonly have scintillation

Exploded view of International Ultraviolet Explorer spacecraft, another observer of high-energy phenomena. A mere 4.3 m long, IUE has a telescope protruding from top (left) and four on-board television cameras to convert UV spectral displays into video signals for transmission to ground. Explorer's range includes distant galaxies and space between them, quasars, pulsars, binary star systems, interstellar dust, as well as fundamental emissions of carbon, hydrogen, nitrogen, and oxygen—the basic constituents of human life on Earth.

counters as sensors.

Fascinating observations of stars, planet-breeding interstellar clouds, and galaxies have been made by spacecraft including the sophisticated International Ultraviolet Explorer (IUE). The fundamental emissions of the common elements of life—hydrogen, carbon, nitrogen, and oxygen—are within the IUE's scope. The fourteen-foot (4.3-m) octagonal IUE has a telescope protruding from the top and a solar array on both sides to provide electricity. Four television cameras on board transform spectral displays into video signals for transmission to the ground.

X-ray astronomy started from rockets in the early 1960's and has mushroomed aboard spacecraft. Shortly afterward, orbiting gamma ray telescopes first made significant discoveries flying Vela satellites. The High Energy Astronomy Observatory (HEAO) is the newest program to study how extremely high energies are generated in space and how the universe evolves.

Three nineteen-foot (6-m) observatories, each weighing 7,000 pounds (3,200 kg), including 3,000 pounds (1,400 kg) of experiments, succes-sively orbit Earth above the screening atmosphere. Images acquired by the telescopes are transmitted to ground stations and reconstructed as pictures showing the size and structure of powerful high energy sources. HEAO 1 scanned the whole sky and quadrupled the X-ray source inventory in 1978. HEAO 2 in 1979 probed especially interesting spots found by its predecessor. HEAO 3 will scan again following launch in 1979. Significantly, HEAO 1 detected great clouds of hot gas that increases the total matter known to exist in the universe. The HEAO data will be analyzed for many years, providing further clarification of the universal processes that lead to life on Earth and possibly elsewhere.

X-rays are typically emitted by hot gas. Gamma rays theoretically are produced in the core of a source by nuclear processes. An orbiting spacecraft called the Gamma Ray Observatory (GRO) is on the drawing board for 1984 to wrest information about cosmic evolution from the most energetic form of radiation known. The 1980's hold great promise for seekers of other planetary systems if we continue along the inspiring and illuminating paths we have discovered.

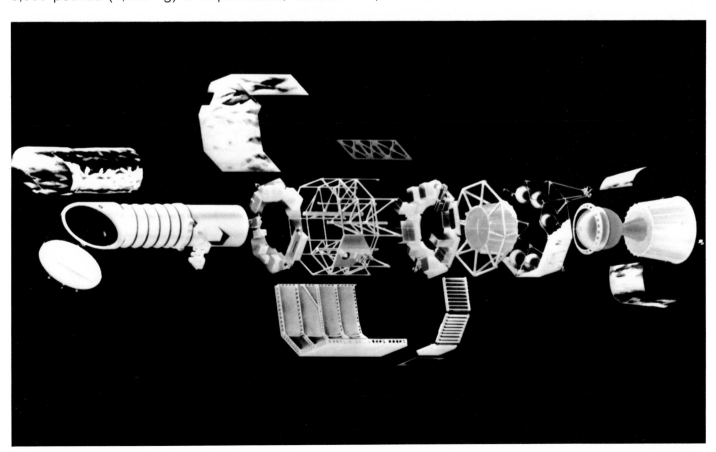

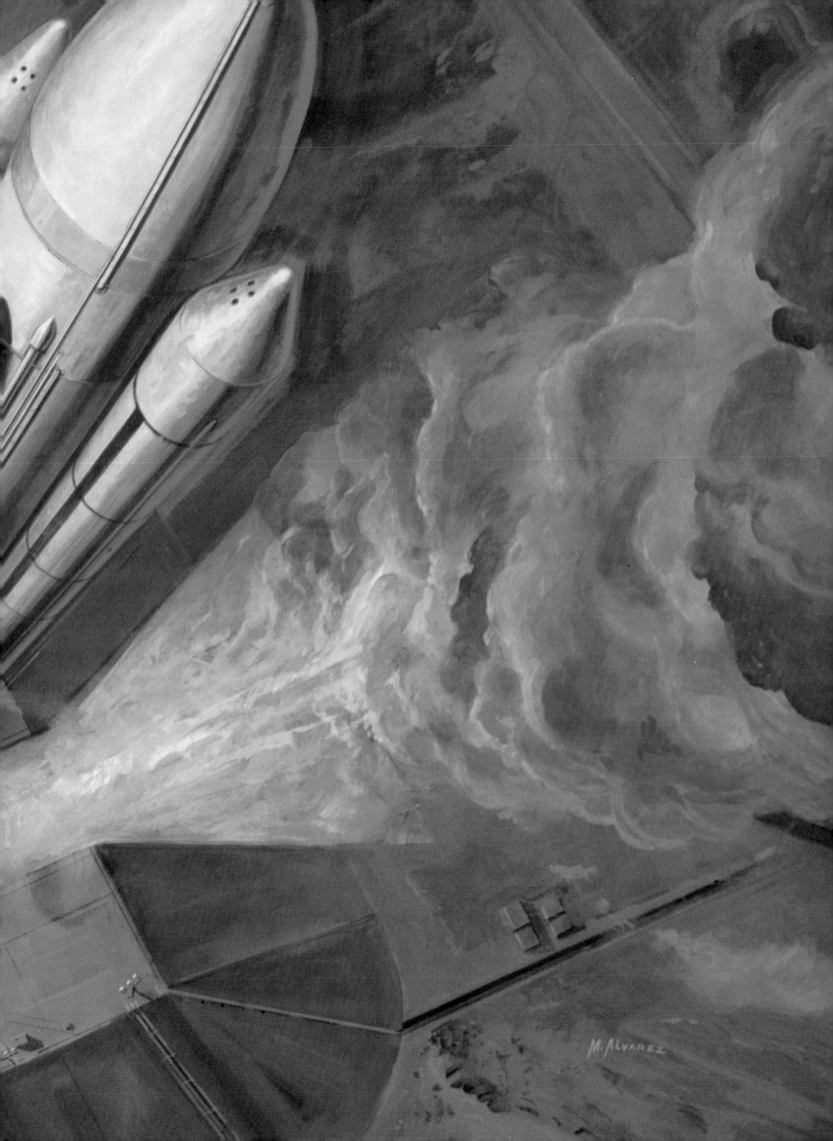

Space Travel

An immense, dark void separates human space travelers from the stars. The difficulties of interstellar travel loom menacingly. Yet our ancestors braved ominous oceans to successfully colonize seemingly inaccessible lands. The human spirit remains dauntless. Theorists and technicians are preparing the way for long space treks. Surely we can expect our descendants to sail to the stars.

The Moon

The record human space trek to date is the 240,000-mile (384,000-km) journey to our moon. Between 1969 and 1972 the Apollo program had twelve space-suited astronauts explore the desolate lunar surface while six others remained in lunar orbit in the Command Modules. All astronauts were carefully monitored before and after each lunar expedition. Their experience opened another world to humans, evidenced no insuperable physical or psychological barriers to human space travel, and supplied data suggesting that life could indeed evolve beyond Earth.

The Apollo astronauts returned 843 pounds (379 kg) of rocks and soil for careful laboratory inspection. They left scientific instruments on the moon that were finally turned off only in 1977. Three automated Soviet Luna spacecraft went to the moon and brought back several more ounces of lunar material between 1970 and 1976. Cameras on orbiters obtained 18,000 photographs. Although data collection has been completed for the present, scientific analysis will continue for many years.

The Apollo astronauts saw no living plant or animal on any visit to the moon. When Neil Armstrong took the first human step on the moon "for all mankind" on July 20, 1969, he entered a barren world. The first Apollo astronauts were quarantined upon their return home to prevent contamination of Earth by moon bugs. That precaution was abandoned after the third Apollo mission because the moon was found to be so sterile. In all the painstaking laboratory analysis of the moon rocks that followed, no living organisms or fossil remains of any kind of plant or animal life have been found.

Further, no water was ever found either flowing freely on the moon or chemically combined in rocks. Water plays an extremely important role in life processes on Earth. Thus, all available evidence points to the inescapable conclusion that life most likely never even started on the moon, although the arid moon does have the critical carbon, oxygen, and nitrogen atoms necessary for life. The lunar soil does have simple molecules formed from these atoms. We now have samples containing the raw materials of life from another world. Conceivably, life might have developed from these simple molecules and more complex ones had the environment been more favorable.

The moon is 2,170 miles (3,470 km) in diameter and only 1/81 as massive as Earth. Because of its small size and mass its gravity is much weaker than Earth's. It cannot hold onto an atmosphere, so there is no air. Everything on the moon—including an astronaut—weighs only one-sixth of what it does on Earth.

Although the moon has no indigenous life, it

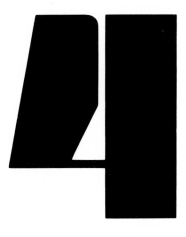

Preceding pages: With main engines and
solid rocket motors roaring, space shuttle
lifts off from launch pad. About size
of a small jetliner, shuttle will be first
reusable space transportation system. Above:
Early conceptions of moon travel. Weightlessness
and weird inhabitants (top & bottom left)
from H. G. Wells' First Men in the Moon. Jules
Verne capsule (top right) flies From Earth to
the Moon through starry sky. Bottom:
Fanciful moonfolk from 1836 English pamphlet.

could serve as a base for humans. While water and air are absent, and deadly radiation bombards the surface relentlessly, these are difficulties that technology overcame for the Apollo astronauts. The Apollo explorations indicate that our inhospitable neighboring world could be made habitable for many people.

The moon is made of igneous rocks formed by the cooling of molten lava. The dark regions that shape the features of the "man in the moon" face we see from Earth were misnamed maria (seas). These are actually three-billion-year-old lava beds similar to those in Hawaii and Iceland. The light-colored regions called highlands are rocks that cooled slowly inside the moon some 4.6 billion years ago.

Moon rocks are made of the same chemical elements as Earth rocks, but in different proportions. The two can be differentiated under a microscope. Maria rocks have more titanium, iron, and magnesium than do most Earth rocks. Highland rocks have more calcium and aluminum. Rarer elements with high melting points, such as hafnium and zirconium, are also more plentiful in lunar rocks. Mining operations could be established to supply future space bases or even an Earth depleted of irreplaceable resources.

Environmental conditions on the moon are strange yet not impossible. With no air to carry sounds, ghostly silence prevails. Days and nights each last as long as fourteen of our days. Surface temperatures range from a stifling 250°F (121°C) at local noon to a frigid −280°F (−138°C) at night. There are numerous circular craters, rugged moun-

Giant Steps: First human landings outside Earth's environment were remarkable Apollo moon missions. These proved feasibility of spaceship design and, with subsequent flights, human adaptability to stresses of zero gravity. While much remains to be learned, it seems certain that space travel will become commonplace. Opposite: Apollo blast-off from Cape Kennedy. Above: View of Earth from Apollo 11 as it emerged from behind moon after lunar insertion orbit burn. Top: Apollo 12, en route to second moon landing, observes Earth eclipsing Sun, totally.

tain ranges, and deep winding canyons called rilles. The landscape is covered by a layer of fine powder resulting from billions of years of continuous bombardment by large and small meteorites.

The Apollo missions demonstrated beyond a doubt that humans can live and work successfully in the moon's hostile environment. There are no present plans to send new astronauts there, but by the twenty-first century lunar bases could be built so humans can resume scientific exploration.

Because it has no air, the moon would be especially suitable for astronomical observatories. Optical astronomers could point their telescopes to the stars without having to struggle against an obscuring veil of air as they must even at the best earthbound sites. The lunar sky is always dark and clear.

No air or water ever paints blue skies, white clouds, or dark storms on the moon. Gamma ray, X-ray, ultraviolet, and infrared observations that are impossible to make beneath Earth's atmosphere could go on continuously there. Geologists could do research on lunar rocks of a nature that is absolutely impossible on Earth because of Earth's water erosion.

The far side of the moon would be ideal as a research site for radio astronomy. On Earth, radio telescopes are very susceptible to interference from local airwave broadcasts. The moon's globe blocks radio signals from Earth. Radio telescopes erected on the moon's far side could listen free of earthly static for messages from extraterrestrials. Antennas might even be built into craters to provide a huge shielded array of "ears" to detect stray sounds.

Future lunar colonists could live inside domes and in underground cities protected from the hazardous lunar environment. Improved flights shuttling between Earth and moon will almost certainly become a reality, but the colony should aim to be ultimately self-supporting. Efficient surface transportation will be developed utilizing experience gained with the early model Apollo lunar rovers. Water and oxygen will likely be processed from surface materials. One eager airline company has already offered flight tickets to the moon. Possibly within our lifetimes tourists will join scientists on visits to our neighbor world. And from launching pads at the future moon bases, they can go on to deep space!

Prelude To Long Space Flights

Space travel subjects humans to unfamiliar environmental hazards. Astronauts confined in close quarters in zero gravity have experienced physical and mental stresses foreign to normal Earth life. Medical monitoring of crews indicates that distinct, though apparently not dangerous or permanent, physiological and psychological changes commonly occur when humans stay in a weightless condition for prolonged periods.

Weightlessness causes changes in blood composition and in circulatory and heart function, reduction of muscle tone, loss of bone calcium, and malfunctioning of the vestibular system of the ear which disturbs orientation—a person's sense of location within an environment—and awareness of

Astronauts' work on moon demonstrated capacity for sustained effort in sterile environment. Top left: Apollo 17 commander Eugene Cernan is photographed near lunar rover. Note tread marks in moon dust. Right: Module pilot Harrison Schmitt uses lunar scoop. Instrument atop boulder is gnomon for measuring altitude by noon shadow it casts. This was sixth landing. Time on moon was record 75 hours. Bottom: Apollo 15— fourth landing—made first use of lunar rover. Astronaut James Irwin drives. Lunar module is at left. Flag is true-color guide.

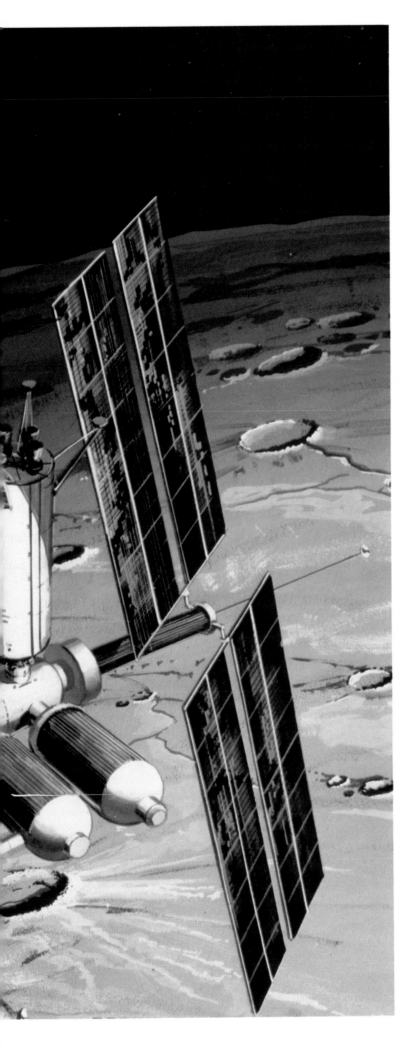

the pull of gravity. The astronauts' superbly trained bodies always readapted quickly to their accustomed environment on Earth with no evident lasting effects. Scientists are now conducting studies to determine how ordinary space travelers can best adapt to life and work in a weightless environment.

Men and women from different age groups and professions are voluntary guinea pigs in tests at the Biomedical Research Division at the NASA Ames Research Center in California. Stringent physical and psychological experiments evaluate human ability to work efficiently aboard spacecraft while in zero gravity, as well as tolerance for the abnormally high gravity and pressures of reentry into Earth's atmosphere. Risks are being calculated and remedies sought to maximize successful human adaptation to space travel.

Bed rest is an excellent physiological approximation to the zero gravity experienced in space and is used to measure bodily responses likely to occur there. Bodily changes occur within a day or two after complete bed rest begins and continue to develop throughout the next week. The bed rest is total and absolute. Volunteers perform all bodily functions flat on their backs.

During bed rest, medical attendants continuously take numerous blood samples, record body temperatures and electrocardiograms, and monitor sleep patterns. Volunteers are permitted to prop their heads slightly on a standard pillow to eat; however, they are wheeled in a horizontal position into a specially developed facility for showers. Nozzles along the sides of the stall provide water that

Orbiting Lunar Station: Hardware for broad range of continuing moon studies and support or control of unmanned orbital and surface operations. OLS would include a space station, lunar transport vehicle, resupply modules, fluid module, and crew transfer vehicles. Nine modules would be required for crew of eight, subsystems, and consumables. Tenth would contain scientific equipment and sensors. Among studies might be analyses of moonrock metals (titanium, magnesium, iron), chemicals (calcium), rare earths (hafnium, zirconium) available for space colonies or depleted Earth.

flows horizontally, somewhat like an automatic car wash.

Volunteers learn to fly a flight simulator while in a sitting position and lying down, since during the bed rest tests they must do everything from a supine position. Test "flights" of the simulator are placed above a subject's bed for the purpose of measuring reflexes and coordination.

A human centrifuge simulates abnormally strong gravity. Volunteers are strapped in a chair that spins at varying speeds similar to an amusement-park ride. They experience a force pushing them down into the chair like that felt in spacecraft at takeoff and reentry. Centrifugal forces range from normal gravity to three times normal, at which point blood is pulled from the head and people feel, or actually do, faint.

So far no volunteer has suffered any serious physical change during the tests. Discomfort and boredom have been the strongest adverse reactions. These experiments indicate that people of all ages will one day travel in space as routinely as they fly in airplanes today!

Space Stations

Space is the best place to determine accurately what physiological and psychological standards are acceptable for flights of long duration. Both the American Skylab and Soviet Salyut 6 space stations have had extensive in-flight research programs to establish health criteria for space treks.

Skylab was the pioneering American space station. Three different crews zoomed to the space station in Apollo spacecraft in 1973-74. Skylab 4 astronauts Gerald P. Carr, Edward G. Gibson, and William R. Pogue set the record for the longest single-manned American space mission—at eighty-four days, one hour, sixteen minutes. The three crews conducted medical experiments to determine how men adapt to weightlessness. Altogether they spent a total of 171 days, thirteen hours, and fourteen minutes in the space station.

Life inside Skylab was a preview of how people will live on long space flights. Some astronauts experienced motion sickness because of weightlessness at the beginning of their flights. Conditions aboard Skylab apparently provoked errors in several experimental procedures initially. Future space flights will have to provide sufficient exercises, more relaxation time, and adequate diets to prevent the effects of prolonged weightlessness and isolation from disabling space travelers.

Daily routines were changed in amusing ways by weightlessness. In zero g, anything not fastened securely will float. Astronauts glided about their space abode like birds. They donned special shoes that locked into the floor when they wanted to do serious work. If someone spilled a glass of water, the liquid did not fall down. Showers were managed with the aid of a vacuum to suck up drifting water drops. When they wanted to get a real rest, astronauts strapped themselves securely into sleeping bags.

Meals were an important diversion within the small confines of Skylab. Almost a ton (.9 t) of food was stowed at launch for the whole mission. There

Top: Soviet Soyuz—meaning "Union"—about to dock with large Salyut 6 space station. A long-range research and reconnaissance facility, Salyut has spent 5 months in Earth orbit. Bottom: Ground control at Kalinin, near Moscow. Map shows orbital complex. At left Salyut is linked at either end with Soyuz 29 and 31 to permit crew transfers. Television picture (right) shows cosmonauts with East German visitor. Salyut is also refueled and resupplied by Progress robot cargo capsules. This means much lighter loads at launch and more room for vital equipment.

were about seventy different frozen and freeze-dried foods plus powdered drinks. Ice cream was most popular; cold potatoes were unanimously hated. The astronauts complained about eating a carefully controlled diet. They recommended that future space travelers be given a wider choice of foods on long journeys.

Early in the mission the men suffered congested sinuses, red and puffy eyes, and swollen faces. Although these flulike symptoms were annoying at first, everyone became used to them. Fingernails and toenails grew more slowly than on Earth, requiring cutting only once a month. The astronauts grew about an inch taller in weightlessness as the vertebrae of the spinal column stretched and body fluids shifted from the lower to the upper extremities.

Astronauts in spacesuits spent a total of ten hours, thirty-two minutes performing extravehicular activities. They made critical repairs and replaced film packs completely outside the spacecraft. They demonstrated that people can successfully accomplish construction and maintenance tasks in space.

Upon returning to Earth, the Skylab astronauts were in better shape and readjusted more quickly to normal gravity than had the Apollo astronauts. Their exercise program was probably responsible for their good health and readjustment. Everyone used a portable treadmill and stationary bicycle up to an hour and a half each day in Skylab to keep in shape.

After studying Skylab data, many doctors agreed that there is no medical reason why people cannot go on extended missions to the planets and beyond, although the shift of body fluids that was probably responsible for the initial motion sickness and some psychological distress should be eliminated on future space flights. Preventive measures are being sought, including further exercises and better diet rather than the use of drugs.

Skylab proved that space stations are especially useful astronomical observatories. Skylab's eight telescopes snapped 182,842 images in the X-ray, ultraviolet, and visible portions of the spectrum, including unprecedented images of our sun and the passing Comet Kohoutek. Skylab also produced 40,286 pictures of planet Earth that give vast quantities of data on natural resources.

Cosmonauts Vladimir Lyakhov and Valery Ryumin set the world's record for human space flight in 1979. They spent over five months in orbit aboard the Soviet Salyut 6 space station. Now long space flights lasting several months seem within our grasp.

The Salyut 6 spacecraft is a manned orbital research and reconnaissance facility. It has hosted cosmonauts from Poland, Czechoslovakia, and East Germany in addition to those from the Soviet Union. Cosmonauts are trained at Star City, near Moscow. They live and work in shirt sleeves aboard Salyut 6 and occasionally bundle up in spacesuits for extravehicular activities. Supplies, including such homey items as garlic and guitars, are ferried up to the space station as necessary by revolutionary unmanned supply capsules called Progress.

Progress craft make intermittent manning

Above: Apollo 18 (left) and Soyuz 19 rendezvous 225 km above Atlantic Ocean for first international manned docking in space. Craft were joined for 89 minutes while crews exchanged visits and meals. Success of mission increased space-rescue capability. Opposite: Life aboard Salyut 6 is enlivened by floating chess (left) and photographing gadgetry. Cosmonauts usually work in shirt sleeves, don spacesuits only for extravehicular tasks. Gear, garlic, and guitars are ferried up as needed. Orbiters are first step toward space colonies.

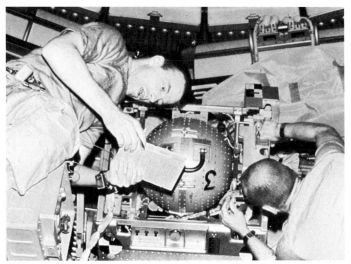

and equipping of space stations newly possible. The first orbital docking of an unmanned Progress cargo capsule with the Salyut 6, and the first refueling of an orbiting space laboratory, were accomplished in 1978. No longer must all supplies, propellants, and scientific apparatus be launched at the start of a mission. These can be replenished by robot capsules leaving room on the space station for the most important items.

The possible future uses of orbiting space stations are unlimited. They would be ideal sites for materials processing and manufacturing research in zero gravity, astronomical observations of deep space, survey platforms for Earth reconnaissance, and launching pads for interstellar flights.

Space Shuttle

By 1979 American astronauts had logged a total of 22,503 hours in space flight. These highly trained men went into space under very special circumstances. Crews of one to three men in superb physical condition flew in expensive spacecraft that could be used only once. Space shuttle turns space flight into a routine, economical possibility for any reasonably healthy men and women who have important work to do in orbit.

The shuttle orbiter takes off like a rocket, orbits the Earth for up to a month, and glides back to Earth like an airplane. It can shuttle to and from space a hundred times or more. The orbiter is about

Skylab: Despite lift-off damage which ripped away vital heat shield and one of two solar panels (opposite top), Skylab goals were achieved. First crew made ingenious repairs, others came aboard in turn, did their chores (bottom), and departed. Above: Straight-down shot through two workshop levels. Space suits are empty. At top right is handle of parasol jury-rigged as heat shield. Hatch behind astronauts is for trash. Left: Lab-treated photograph of Comet Kohoutek taken from Skylab 4. Tail is 4.8 million km long, shows dust and gases blown from comet.

the size of a small jetliner but is much more complex. It carries both crew and cargo. The orbiter is 122 feet (37 m) long and fifty-seven feet (17 m) high, with a wingspan of seventy-eight feet (23 m). It weighs about 150,000 pounds (68,000 kg).

The 88,000-acre (35,000-ha) Kennedy Space Center at Cape Canaveral was modernized to serve as a launching and landing site in the new shuttle era. A new hanger called the Orbiter Processing Facility was erected. A 15,000-foot (4,500-m) runway with 1,000-foot (300-m) overruns at each end was constructed. New communications and computerized check-out systems were installed.

Launch Complex 39 and the Vehicle Assembly Building, which captured world-wide attention during the Apollo and Skylab programs, were both modified for the shuttles. Towers for launch are fixed at each launch pad. Payloads as large as 65,000 pounds (29,500 kg) can be launched due east from the Kennedy Space Center into an orbit of 28.5 degrees inclination.

Shuttles transporting U.S. Defense Department payloads, such as secret reconnaissance satellites, will operate out of the Vandenberg Air Force Base in California. Payloads of 32,000 pounds (14,500 kg) can be launched from the Vandenberg Air Force Base into an orbit as high as 104 degrees inclination. Polar orbiting capabilities up to 40,000 pounds (18,000 kg) can be achieved from there.

In a typical mission, the orbiter's main engines and the boosters ignite simultaneously to rocket the shuttle into orbit. In space, two unmanned, solid rocket boosters separate from the orbiter and parachute to sea for recovery and reuse. A large external propellant tank is jettisoned just before the shuttle starts orbiting. The external tank breaks up in the atmosphere over ocean area.

The crew controls launch, orbital maneuvering, atmospheric entry, and landing phases of the mission from an upper level flight deck at the forward end of the orbiter. Pilot and copilot sit side by side in a cabin resembling that in a jetliner. A mission specialist, who manages equipment and supporting resources, and a payload expert can sit behind them. These people are NASA astronauts. Seating for six passengers and a living area are on a lower deck. Payload experts who conduct the mission's experiments do not have to be astronauts to do the work effectively.

People experience three times normal gravity during the shuttle launch but less than 1.5 g during a typical reentry. They wear ordinary clothes and breathe normal air that is composed of 20 percent oxygen and 80 percent nitrogen. The food for the space shuttle is developed and tested in an ultra-modern kitchen at the NASA Johnson Space Center in Texas. Seventy-four kinds of food and twenty different beverages are available. The calorie count is 3,000 daily, since working astronauts need the same caloric intake to survive as do people on the ground.

Different food-preservation techniques are employed to maximize the variety of meals available aboard space shuttles. Menus have heat-processed foods canned or packed in laminated foil pouches, such as beef with barbecue sauce and

Top: Scientist-astronaut Owen Garriott, operating outside Skylab 3, deploys Particle Collection S149 Experiment on one of four Apollo Telescope Mount panels. Purpose is to study impact of interplanetary dust particles. Far right: Bakersfield, CA, area. Skylab photos and Earth sensors monitored snow melt, cropland, weather, health of forests, possible new ore deposits. Right: Skylab flew 2,476 times around globe—some 112 million km—with crews aboard. Mission was scientifically productive and proved long-term space flight feasible.

tuna fish. Foods preserved by controlling the available moisture are dried apricots, peaches, and breakfast bars. Dehydrated foods that can be reconstituted with water include scrambled eggs, beef patty, chicken and noodles, and all beverages. Some foods, such as bread, rolls, and beefsteak, are exposed to ionizing radiation to preserve them. Finally there are foods such as nuts and cookies in their natural form.

The crew should be able to get a meal ready to eat in about an hour. The shuttle galley has a pantry, an oven to heat meals up to 185°F (85°C), a unit for washing up, hot-and-cold running water, and a dining table. Trays and utensils are cleaned with wet wipes containing a cleaning compound.

Astronauts might have to go outside the orbiter to inspect, operate, or repair equipment, for photography, or to assemble or replace modular materials. Unisex spacesuits are ready for them in small, medium, and large sizes to adjust to all pilots and mission specialists. An airlock and mobility aids for weightlessness, such as handrails and foot restraints, are also available so spacesuited astronauts can do extravehicular tasks without unnecessary difficulty.

If there is an emergency that disables the shuttle, a rescue orbiter can be launched to transfer astronauts and passengers from the marooned craft. Each passenger nestles inside a 34-inch-(86-cm) diameter ball called a personal rescue system, which provides a short term life-supporting environment. The ball is made in three layers of Urethane, Kevlar, and an outside thermal protective layer. It

has a small, tough, thermoplastic transparent viewing port and its own communication system.

The same materials are used in the spacesuit. A strong and durable Kevlar fabric makes better mobility possible for elbow, wrist, and knee joints, at reasonable cost and weight. The life-support system is an integral part of the rigid upper torso of the shuttle suit. Passengers in their personal rescue balls are transferred by spacesuited astronauts from the disabled shuttle to the rescue orbiter.

Potential uses of space shuttles to improve life on Earth and extend life into space are unbounded. One shuttle can ferry five robot satellites at a time up to orbit, facilitate maintenance of others already there, and return disabled craft to Earth. The robot satellites include those used for agricultural surveys, communications, defense, environmental protection, energy, mapping, navigation, oceanography, and weather forecasting.

Each orbiter has a long mechanical arm to manipulate payloads outside the craft. A second arm can be installed when two manipulators are required for handling payloads. Each arm has lights to provide illumination and remotely controlled television for side viewing and depth perception. The arm can be used for jobs such as placing robot satellites into orbit or retrieving them for repair.

Spacelab is a complete scientific laboratory adapted to operate in zero gravity in orbit aboard the shuttle. Scientists work there in ordinary clothes in a normal environment. They are reasonably healthy men and women from many nations who are experts in their fields and fly in Spacelab after only

Top: Satellite power system being assembled in space to convert Sun's radiance into electricity for energy-hungry world. Spindly grid of lightweight parts will be adequate in no-stress environment of zero-gravity space. Crews, materials, and equipment will be transported by shuttles, like that being unloaded (upside down) atop main station. Bottom: Orbiter with multiple payload. One satellite has been released, second is about to follow. Advanced radio-astronomy unit lies aft for deployment elsewhere.

a few weeks of rigorous space-flight training.

Experiments look for the effects and possible utilization of weightlessness in medicine and manufacturing. Drugs of unusual purity, new metal alloys, glass, and electronic crystals can apparently be manufactured best in a weightless environment. Astronomical observations unhindered by the atmosphere should reveal new wonders. The Spacelab can be removed from the shuttle for reoutfitting and be reused about fifty times.

Spacelab provides an ideal observation platform for monitoring crisis situations on Earth, such as sudden stormy weather, industrial accidents that release pollutants, and crop blights. The shuttle whizzes over the United States in eight minutes!

After each mission, the shuttles return to Earth for overhaul and refurbishing. The orbiter is protected from the searing heat of reentry into the atmosphere by semipermanent silica-coated ceramic blocks that radiate heat back outward. It glides to a landing, decelerating from an orbital speed of 17,000 miles (27,000 km) per hour to a landing speed of about 220 miles (350 km) per hour.

In the future, space shuttles could carry into orbit the building modules for large solar power stations to convert abundant sunlight into unlimited supplies of electricity for an energy-hungry world. They could also carry modules to construct orbiting space colonies. The materials would be assembled by specialists who are transported and supported by a shuttle. Orbiting shuttles could serve as launching sites for accelerating spacecraft into deep space.

Interstellar Travel

Perhaps humans will follow the robot Pioneers and Voyagers across the boundary of the solar system in the next millennium. It is true that an enormous vacuum separates Earth from the nearest stars. But humans have successfully navigated seemingly impossible spaces before. Historically, human ingenuity in the face of great challenges has been awesome.

This time the problems are truly stupendous, although probably not insurmountable. The closest visible star, Alpha Centauri, is 4.3 light-years away. At conventional rocket speeds, space travelers would require 40,000 years for a one-way trip there. A journey to one of the possible planets circling Barnard's Star would take 100,000 years. Since the other billions of stars in our Milky Way galaxy are even farther out, travel times would be longer still.

Distances to the stars are immense, and human life spans are miniscule on a cosmic time scale. Today star treks are not possible, but they could become feasible with future increases in scientific knowledge and technological capability.

One possibility is radically improved starship propulsion systems to replace conventional ones. The fastest starship conceivable would have to be propelled at practically the speed of light, or 670 million miles (1,080 million km) per hour. (Einstein's Theory of Relativity sets the speed of light as the speed limit of the universe.) There is no practical method of attaining speeds anywhere near that of light today, but numerous innovative designs are

Opposite: Shuttle (top) sheds its solid rocket boosters at 5,166 km/hr. SRB's will descend on parachutes, be recovered, refurbished, and reused. Big external fuel tank will be jettisoned later. Bottom: Cutaway shows shuttle configuration carrying European Space Agency's Spacelab. Instruments will be exposed to zero gravity. Left: Glowing with heat of re-entry, shuttle returns to Earth atmosphere after mission in space. Exterior has silica-coated ceramic tiles that radiate heat outward. From orbital speed, shuttle will land at modest 350 km/hr.

Innovative film **2001: A Space Odyssey**
considered many problems of space travel. Top:
To counteract zero gravity on long voyage
to Jupiter, spaceship control room
rotates slowly, providing centrifugal force
and sensation of weight for crew along
walkway at rim. From any point, "up" is toward
hub. Life processes of other astronauts
are suspended by state of artificial hibernation.
Medical monitoring is automatic. Below:
Weightless stewardess is en route to overhead
cockpit area to serve meal to flight crew.
This is shuttle flight to space station.

being suggested for future light-speed travel.

One theoretical propulsion system would utilize the same hydrogen fusion as occurs in the cores of stars and in nuclear H-bombs. The moving starship would scoop up hydrogen from space and harness all the released energy from hydrogen fusion to propel itself forward at light speed. Another theoretical propulsion system would utilize complete matter annihilation to release tremendous energy. When minute amounts of matter and antimatter are brought into close contact in a laboratory, they annihilate each other. Matter and antimatter disappear, leaving a flash of pure energy. If large quantities of matter could be combined in a controlled way with antimatter, the starship could theoretically be pushed to light speed. The technological problems in achieving speeds approaching that of light are formidable, so it will be some time before any such starship is actually built.

If a starship that zooms at light speed becomes possible, time will be less of a problem for future space travelers than it is today. Special relativity effects, described by Einstein, permit light-speed travelers to age much more slowly than less adventurous folk who remain behind on Earth. Time as measured from Earth seems to slow down on starships traveling at light speed. During a century clocked on Earth, light-speed travelers would age only about three years.

Deep-freeze storage offers another possibility for future star treks. Today animal and human sperm are successfully frozen and unthawed for artificial insemination. Normal human births have occurred after insemination by sperm frozen at least ten years. Human fertilization was first accomplished in a test tube by Dr. Patrick C. Steptoe in 1978. Perhaps frozen sperm and egg cells will be placed aboard future starships. These cells might be implanted in an artificial womb and combined under computer control when the starship neared its destination.

Viable mice, rabbits, sheep, and cattle have emerged from embryos that had been thawed and transferred into foster-mother wombs. Conceivably human embryos could be frozen and stored close to absolute zero aboard starships. They might be artificially nurtured to birth under computer control centuries later.

If cryobiologists ever succeed in freezing adult bodies for future thawing, passengers could spend hundreds of years in suspended animation in a deep freeze aboard a computer-controlled starship. Upon arrival at the predetermined destination, the space travelers could be awakened by the computer to venture forth on new worlds.

A very different alternative for interstellar travel was proposed by physicist Freeman Dyson of Princeton University. He calculated that in two hundred years a spacecraft reminiscent of Noah's Ark could be built to waft people to the stars in a journey of a few centuries. The ark would be a self-sustaining colony with plants, animals, and people. Generations would be born, live complete lives, and die en route to the stars. If humans accept efficiency and cost-effectiveness in place of super speeds, such a space ark could be the first to embark.

Search for Extraterrestrial Intelligence

A systematic search for intelligent extraterrestrials in the Milky Way galaxy is within the capacities of exobiologists using available contemporary techniques. Pioneers have listened sporadically for messages from outer space ever since physicists Guiseppe Cocconi and Philip Morrison proposed a feasible search strategy in 1959. Many of the early experimental problems have been solved. Highly respected scientists have participated in numerous significant discussions of search purposes and techniques. They agree that a serious search for extraterrestrial intelligence would be productive regardless of its ultimate outcome.

Probabilities

Those supporting a serious search believe there is a good chance that other galactic civilizations exist and desire to communicate. Astronomers Frank Drake and Carl Sagan of the United States and I. S. Shlovsky of the Soviet Union listed conditions that must be satisfied if the search is to be successful in terms of an impressive equation. Seven factors multiplied together yield an estimate of the number of civilizations capable of communicating with Earth at this point in galactic history. All the numbers plugged into the equation are guesses. Nevertheless, the method provides useful insights and is widely used.

The primary assumption is that, like us, any other galactic civilization must inhabit a planet circling a star that can sustain it. The first three factors in the equation are estimates of 1) the rate at which stars form in our galaxy every year, 2) the fraction of the stars that have planets, and 3) the number of those planets where environmental conditions are suitable for life to develop.

Thereafter, the assumption is that unicellular life emerges elsewhere and that it subsequently ascends to form intelligent societies that desire intergalactic contacts. The next three factors in the equation are estimates of 4) the fraction of suitable planets on which life actually arises, 5) the fraction of planets where simple life evolves to intelligent creatures, and 6) the fraction of those civilizations that develop both the capability and desire to communicate with others in the galaxy.

The last factor, 7) the lifetime of an intelligent communicative civilization, is the most enigmatic of all. When creatures reach the stage of development we have, do they survive for a few decades and then self-destruct in a nuclear or other holocaust? Or can intelligent civilizations harness technology constructively and survive for millions or billions of years? No one knows.

Each of the successive factors is uncertain. The last figure varies widely with personal philosophical outlook. When the seven factors are multiplied together, estimates range from a pessimistic one intelligent civilization (Homo sapiens) to an optimistic fifty million possibilities in our Milky Way galaxy!

Interstellar Communication

Humans cannot currently traverse the vast distances that separate Earth from neighbor stars. Even if extraterrestrials live relatively nearby, we

Preceding pages: Theirs or Ours. Artist Edward A. Anderson's vision of interstellar warcraft at speed. Esoteric gun turrets are atop vertical stabilizer and cockpit. Human eagerness to find friends on other worlds is tempered by fear that intelligent aliens may be technologically superior and hostile. Above: Three Arecibo-type radio telescopes could be installed in natural craters on far side of Moon for sending radar signals and receiving celestial radio waves.

cannot find them today by sending spaceships to explore their planets. Thus they must act to make themselves perceptible across the void.

Communication among galactic civilizations appears most feasible by means of radio signals. Radio waves zoom through interstellar space at light speed. They pass through huge nebulas and planetary atmospheres that block infrared, visible, and ultraviolet radiation. Humans have achieved the capability of sending and receiving galactic radio messages after a few million years of evolution. If we wanted to broadcast our presence to the galaxy, we would transmit radio messages. It seems likely that communicative extraterrestrials would do the same.

Radio waves are invisible and inaudible. They cannot be recorded on photographic film. Radio telescopes have large metallic collectors to intercept and focus celestial radio signals as giant mirrors do for light waves. Radio receivers detect and magnify ensuing electrical changes. These may be recorded by a pen writing on a slowly revolving paper drum or turned into a hissing noise by a loudspeaker. Significant data may be computer processed into a visible portrayal of the radio source called a radiograph.

Electrical engineer Karl Jansky of the Bell Telephone Laboratories in New Jersey accidentally made the first discovery of radio emissions from space in 1932 while seeking the source of static in transcontinental telephone communications. Since then radio astronomers have detected thousands of cosmic radio sources. Planets, stars, galaxies, and quasars emit a chaotic mixture of radio frequencies without any sensible pattern.

Natural radio signals bombard Earth continuously from many parts of space. They trace distant regions of the Milky Way galaxy that cannot be seen behind huge nebulas. Powerful radio emitters illuminate the brink of the observable universe some fifteen billion light-years away. All celestial radio waves that have been observed appear to have natural rather than intelligent originators. Presumably, if intelligent creatures were trying to contact Earth they would broadcast a pattern such as dots and dashes to distinguish their beacon from natural radio static, but no code has yet been detected.

Radio telescopes must have huge collecting areas because only a few watts of power from cosmic radio sources reach Earth. The world's largest radio telescope is the 1000-foot (305-m) Arecibo radio telescope in northern Puerto Rico. The white concrete, steel, and aluminum structure is built into a natural limestone bowl in the mountains. Both the surrounding hills and the remote setting in a green jungle minimize radio interference.

The Arecibo telescope can intercept celestial radio waves and also bounce its own radar signals off nearby planets and satellites. Its tremendous collecting dish sits immobile in the ground. The reflector surface has an area of twenty acres (8 hectares). It is built of 38,778 aluminum panels that weigh 691,000 pounds (311,000 kg). Natural vegetation in the area prevents erosion and slippage.

Receiving and transmitting equipment that can be steered and pointed by remote control are

Very Large Array (VLA) is world's most powerful radio telescope installation. Capable of detecting purposeful extraterrestrial signals throughout galaxy, it links 27 antennas into equivalent of single superscope 27 km in diameter. Top: V part of Y-shaped network on Plains of San Augustin, west of Socorro, NM. Antennas move simultaneously counter to Earth's rotation, scanning sky from many angles. Bottom: Self-propelled red transporter totes Antenna #2 along track. Spurs off main lines lead to 72 observation stations. #1 (rear) is at one of them.

suspended high above the dish. Antennas hang down from a white steel, triangular support structure. The longest is ninety-six feet (29 m) from base to top. It can receive incoming radio signals or direct signals down to the reflector for transmission to space. The observatory is designed so everything can be operated by just one experimenter.

The largest fully steerable radio reflector on Earth is the 328-foot (98-m) dish at the Effelberg Radiotelescope Observatory in Bonn, West Germany. The largest steerable dish in the United States is a 300-foot (90-m) radio telescope at the National Radio Astronomy Observatory in Green Bank, West Virginia.

As with optical telescopes, more information can be intercepted by ever larger radio collectors. Astronomer Sir Martin Ryle won the 1974 Nobel Prize in physics for developing an ingenious method called aperture synthesis to increase effective collector size. Aperture synthesis combines observations from two or more radio telescopes linked to a computer. It pinpoints the sources of images as well as one king-size antenna would. Very Long Baseline Interferometry (VLBI) is the electronic linkage of two or more radio telescopes separated by great distances to imitate the performance of a single gigantic collecting dish.

NASA's Deep Space Network strategically locates radio telescopes on three continents. Stations at Goldstone in California, Madrid in Spain, and Canberra in Australia are used for VLBI activities. Each has a 210-foot (64-m) antenna and two eighty-five-foot (26-m) antennas. All have transmit-

Fantasies about extraterrestrial beings and other-worldly civilizations were fairly primitive until widening knowledge of the universe increased possibilities of their existence. Now sci-fi books, movies and TV, and approaching reality of space travel encourage speculation about What's Out There, or What's Coming Here. Manifestations include (from left) **Star Wars'** *sinister little* **Sand People,** *slit-eyed* **Invaders from Mars,** *bizarre ruler of* **Cylon** *and armored humanoids from* **Battlestar Galactica,** *and wan, bubbleheaded* **Man from Planet X.**

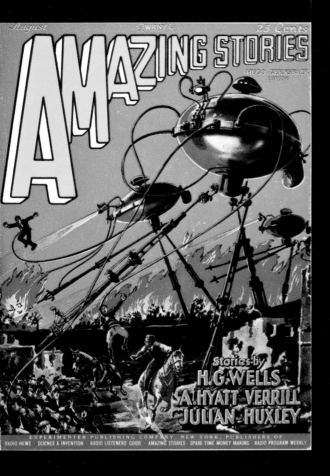

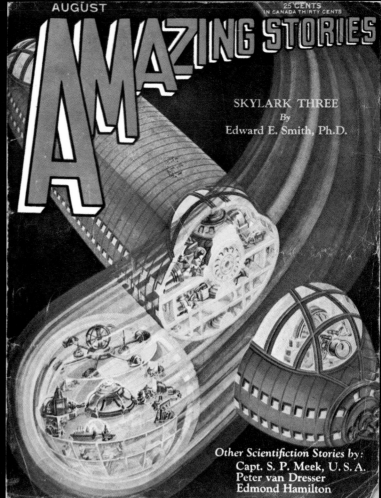

ting, receiving, data handling, and interstation communication equipment. The control center is located at the Jet Propulsion Laboratory in California.

The Deep Space Network has provided tracking, transmission of instructions, and data acquisition support for the Apollo moon explorations and missions to Mars, Venus, Mercury, and Jupiter. It works the long-duration Voyager 1 and 2 flights and may even be able to track the Voyagers beyond the solar system. Existing facilities could be used to search for extraterrestrial intelligence.

The Very Large Array (VLA) is the world's most powerful radio telescope installation. It has the potential to detect purposeful extraterrestrial signals from around the galaxy. The VLA combines twenty-seven fully steerable radio telescopes into the performance of a single superscope seventeen miles (27 km) in diameter through aperture synthesis. Located in the Plains of San Augustin, west of Socorro, New Mexico, the VLA is the primary facility of the National Radio Astronomy Observatory.

Only sparse grasses grow on the Plains. The VLA site is in an isolated dry valley at an elevation of 8,000 feet (2,400 m). Solar radiation beats down mercilessly. Lightning strikes frequently in the surrounding hills. The stark white eighty-two-foot (25-m) antennas appear monolithic above the desolate Plains. They move simultaneously, counter to the Earth's rotation, to scan the sky.

Antennas are distributed along railroad tracks in a "Y" shape. Two arms of the "Y" are thirteen miles (21 km) long and the third, which points north, is twelve miles (19 km) long. Spur tracks perpendicular to the main lines lead off to seventy-two separate observation stations on which the antennas can be placed.

Different views of the sky for mapping or detailed scanning of especially interesting sources are obtained by moving the antennas among the various observing stations in a multiplicity of patterns. A sixty-five-ton (59-t) self-propelled transporter rolls under an antenna, lifts its 214-ton (193-t) assembly off its three piers, and carries it from station to station.

A complex computer system controls the antennas, processes and displays observed data, and produces visible maps and brilliant false-color pictures. When fully operational, the VLA should produce radiographs as sharp as the visible images photographed by the Hale telescope.

Benefits and Risks

A serious search for extraterrestrials must have extraordinary consequences for humanity whether it ends in success or failure. Multiple potential benefits are enumerated by search proponents. Terrifying scenarios are depicted by opponents. Fairness dictates that both benefits and risks be examined prior to inaugurating a major program.

Since publication of his classic 1959 paper, Morrison has frequently pointed out how culturally enriching contact with a different, more advanced galactic society could be for humans. It might permit us a view of possible futures for humanity beyond present crises. Advanced extraterrestrials might teach humans how to control sophisticated

Stilt-legged metallic monsters with ray guns and whiplash tentacles spread panic in H. G. Wells' famous **War of the Worlds.** *Winsor McCay's placid* **Little Nemo** *has beautifully drawn airship journey to Mars. "Canals," a puzzlement until Viking pictures, turn out to be signs of urban sprawl. Multicolored Martians are winged. Atmosphere is suffocating for Earthlings. Space adventure frequently involves inexplicable phenomena launched by extraterrestrials who are mischievous or worse—as on cover of* **Amazing Stories** *in 1930.*

technology for benefit rather than self-destruction.

A message transmitted specifically for the purpose of attracting the attention of less developed galactic civilizations would have to come from extraterrestrials who were convinced that the effort, time, and expense involved were worthwhile. Perhaps they are near-immortals, so that the thousands of years of waiting time required for an answer to their beacon does not appear overwhelming. Perhaps they have a cost-effective means of rapid interstellar travel still undreamed-of on Earth.

Other galactic civilizations could be much older and erudite than ours. How enlightening it would be to link with a vast chain of rich galactic cultures! Anthropologists, artists, lawyers, politicians, philosophers, scientists, and theologians point out that venerable cosmic thinkers could enlarge our deepest personal and social values. Hints from others could lead to big technological gains on Earth. Various scientific breakthroughs, ranging from a valid picture of the evolution of the Universe to whole new biologies, have been imagined. Advanced extraterrestrials could even offer Earth a kind of foreign aid.

Certainly the information in any purposeful message from extraterrestrials must be significant. A deliberate signal should not be too difficult to decode. Senders should design the signal to easily reveal its own code and meaning. If the message is radioed, as anticipated, both the transmitting and receiving societies will share a common understanding of radiophysics. Mathematics and the laws of nature are universal. They provide an intergalactic language that all advanced civilizations should understand.

A major search effort that failed to detect extraterrestrials would still benefit humans in manifold ways. Such an effort would best be organized on a global scale. Necessary international cooperation and enthusiasm could lessen historical tensions between countries. A large-scale program would require increased scientific knowledge and technology. Exciting innovations will almost certainly evolve to meet future challenges.

A lurking fear of star wars haunts opponents of intergalactic communications. It is not impossible that extraterrestrials would colonize and exploit Earth as soon as they discovered it. Nobel Laureate Sir Martin Ryle and others have urged that no deliberate messages be sent out from Earth toward possible galactic civilizations to avoid such danger.

Only one intentional message has been radioed from Earth to space. To celebrate the upgrading and rededication of the Arecibo radio telescope in 1974, the staff of the National Astronomy and Ionosphere Center constructed a coded signal. The Arecibo message was broadcast toward a globular cluster of some 30,000 stars designated M13, in the constellation Hercules. Since M13 is 24,000 light-years away from Earth, the minimum radio response time is 48,000 years.

The Arecibo message opens with a lesson explaining its mathematical codes. It then portrays a human being and human chemistry, and gives information about the Earth's population, its position in the solar system, and its largest radio telescope.

Project Cyclops would assemble 1,000 to 2,500 radio telescope antennas (top) to listen for coded microwave signals from planets up to 1,000 light-years from Earth. Search probably would be conducted in frequency range of 1420 megahertz, where cosmic background noise is minimal and natural radio signal is emitted by most abundant element in universe, hydrogen. Bottom: Ground-level view of 100 m/D antennas. Building at right, in center of array, is data collection and processing facility awaiting first coherent signal.

The message consists of 1679 consecutive bits of information that were transmitted in 169 seconds. After five hours and twenty minutes it zoomed past Pluto and out of the solar system. It is detectable by antennas like the Arecibo telescope anywhere in the Milky Way galaxy.

The spectre of invasions from space has prevented any other deliberate broadcasts from Earth. However, there is no apparent risk in simply receiving messages from space. If humans do not answer, the transmitting society will not even know that its message was received and understood by intelligent beings on tiny planet Earth. We are fairly securely sheltered by great distances and long travel times. Receiving and decoding a radio message from the depths of space promises philosophical and practical benefits rather than peril for humanity.

Strategy

Assuming that there are other intelligent beings in the Milky Way galaxy and that it is productive to search for them, exobiologists talk strategy. Radio astronomy appears most promising today. The search for the thousands of potential inhabited planets by manned starships or swarms of robot probes belongs to the future.

Surmising that first contact will be by radio, astronomers must choose a frequency band in the radio spectrum to tune in. They cannot survey the entire sky continuously in all directions at all radio frequencies. Radio astronomers aim receivers selectively and choose promising channels.

The familiar AM radio band ranges from 540 to 1600 kilohertz while the FM band ranges from eighty-eight to 108 megahertz. (Hertz is the international unit of frequency, equal to one cycle per second. Kilohertz and megahertz mean a thousand and a million cycles per second, respectively.) Besides the world-wide enjoyers of AM, FM, CB, and television, other users of radio frequencies include police radio, radar, military, and satellite communications systems. These frequencies are controlled under national and international arrangements.

A radio frequency band width of six megahertz is allotted to each television channel in the United States. The band width for AM stations is only 1/1000 as wide, and even narrower bands are employed in space communications. Hence, a television set tuned to Channel 4 at a frequency of sixty-nine megahertz does not pick up other channels that are broadcasting simultaneously.

When radio astronomers tune in specific frequencies for possible space messages, they assume intelligent extraterrestrials reason as humans do. Radio signals for interstellar communication must cross immense distances without being absorbed or distorted. They must be capable of detection upon arrival at foreign telescopes. All working radio telescopes pick up a lot of natural background static over much of the astronomical radio spectrum. There is always the microwave radiation left over from the big bang, as well as noises from electrical charges spiraling around in the galaxy. Cosmic background noise is known to be minimal at a frequency of 1420 megahertz.

Radio frequency 1420 megahertz for galactic

Top: Largest steerable reflector in U.S. is 90-m radio telescope at National Radio Astronomy Observatory at Green Bank, WV.
Bottom: Built into a natural limestone bowl in mountains of northern Puerto Rico, 305-m Arecibo radio telescope is world's largest. It can intercept celestial radio waves and bounce radar signals off nearby planets and satellites. Aluminum reflector surface covers 8 hectares—a huge collecting area needed because so few watts from cosmic radio sources reach Earth. Surrounding hills minimize radio interference.

broadcasts has gained many advocates since it was initially proposed by Cocconi and Morrison in 1959. A natural radio signal is emitted at 1420 megahertz by the most abundant atom in the universe, hydrogen. The frequency is not overwhelmed by background noise. It is relatively free of absorption by clouds. Any technologically sophisticated extraterrestrials who were desirous of intergalactic radio communication would likely be aware of these critical properties.

The 1420 megahertz hydrogen signal is just one of dozens of natural microwave emissions from a wide variety of interstellar molecules. Radio frequency 1727 megahertz is a characteristic emission from the hydroxyl (OH) radical. Hydrogen combined with the hydroxyl radical forms water (H_2O). The radio band between 1420 and 1727 megahertz is nicknamed the "water hole" by romantics. They suggest that the first intergalactic encounter will occur at this cosmic water hole just as different animal species meet at water holes on Earth.

A program called Communication with Extraterrestrial Intelligence (CETI) is operating in the Soviet Union. The Soviets are listening for very short radio pulses from outer space. In 1979 N. S. Kardashev of the Soviet Space Research Institute in Moscow proposed monitoring the frequency 203,384.9 megahertz. This is relatively free of noise and of absorption and scattering by clouds. It is emitted by positronium, a short-lived combination of an electron and proton.

Whatever radio frequency the extraterrestrials use, their beacon will presumably be modulated

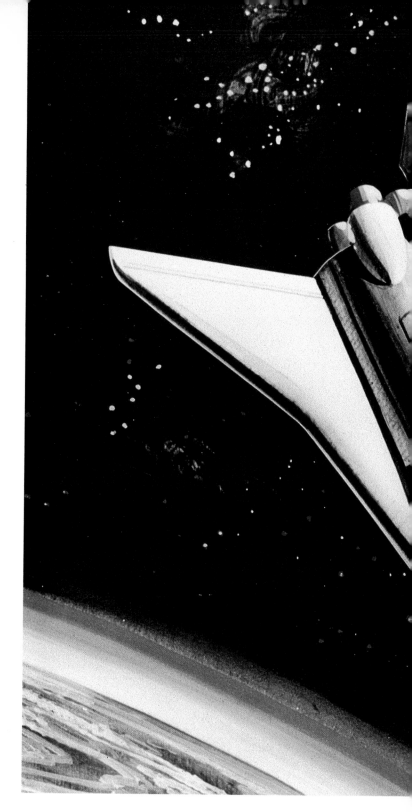

To surmount persistent problem of terrestrial radio interference, lightweight antennas may be lofted into space. Top: Small (30-m) parabolic antenna is assembled and positioned in low orbit by one shuttle flight. Extravehicular astronaut perches on rim. Bottom: Astronaut performs construction task on very large SETI-system antenna in geosynchronous orbit or beyond. Second astronaut and shuttle in background cast shadows on disk. At top are two antennas already in place. Note astronaut's multiple-choice wrench.

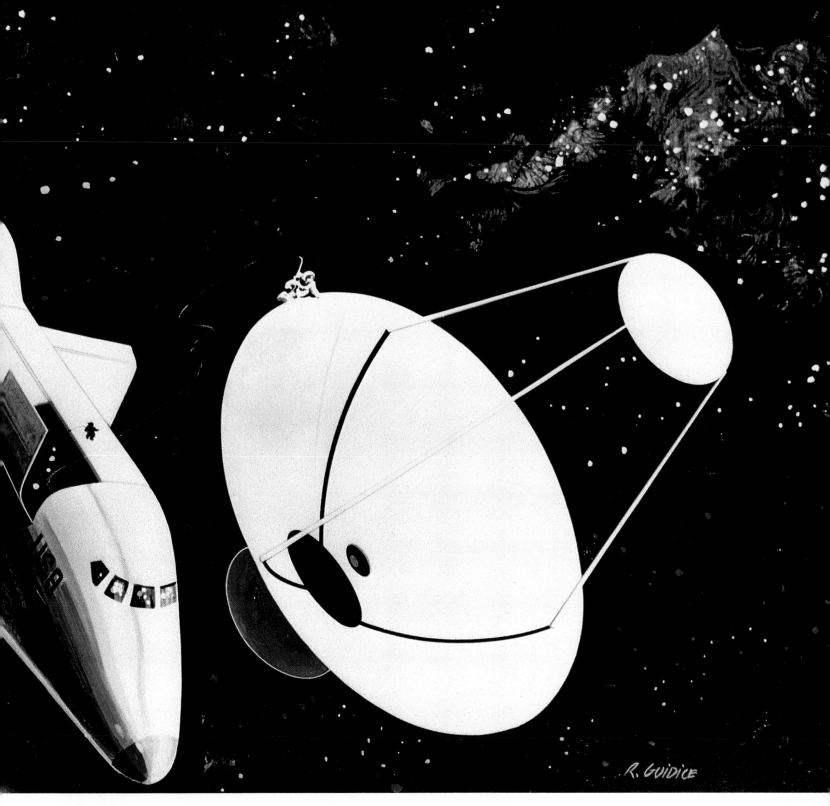

with information content that will in some way distinguish it markedly from background noise. Familiar natural radio signals are not modulated. In addition, extraterrestrials presumably broadcast as we do on narrow frequency bands to optimize their transmitter power. Natural signals characteristically occupy a frequency band that is at least one kilohertz wide.

A few vanguard radio astronomers have snatched brief intervals of telescope research time to listen for purposeful extraterrestrial signals. Their lack of success was virtually inevitable and is not at all discouraging. Not one had the good fortune to make a truly comprehensive survey of the sky.

Astronomer Frank Drake conducted the first search at the National Radio Astronomy Observatory in Green Bank, West Virginia, in 1960. He dubbed the study Project Ozma after the ruler of the exotic land of Oz in the L. Frank Baum stories. The eighty-five-foot (26-m) telescope scanned two stars like our sun—Tau Ceti and Epsilon Eridani—at 1420 megahertz for four weeks. Both stars are only about twelve light-years from Earth. After a month Drake took a break. Then for another month he scanned the whole range of plausible hydrogen frequencies for both stars. Finally, after 150 hours, no more telescope time was available.

The great Arecibo telescope with its 1008-channel receiver can duplicate everything that Drake did in less than a second. Even if Drake scanned correct stars at the right frequency, the extraterrestrial transmitter could have been temporarily turned off. The time and effort devoted to Proj-

Another view of very large SETI antenna with two companions in distance at left. It is a two-feed spacecraft with relay satellite (upper right) and radio-frequency interference shield casting shadow on parabola. So far, search has been for intentional signals, yet even pickup of random radio leakage would establish existence of alien intelligence. But sky is vast. "Even if there are a million space messages coming our way," author says, "radio receivers would have to survey 100,000 stars to enjoy a fair statistical chance of detecting a single one."

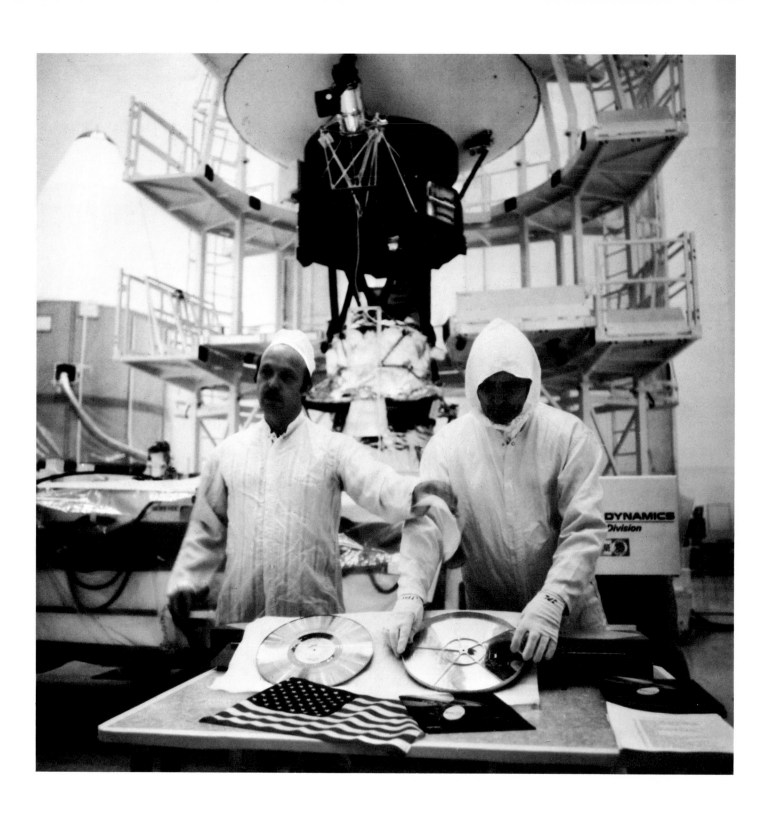

ect Ozma were miniscule compared to that required for a reasonably good chance of successful detection. Drake's counterparts on planets circling Tau Ceti or Epsilon Eridani would likewise have missed Earth if they had relied on an effort similar to his.

Since 1960 radio searches at the Gorky Institute in the U.S.S.R. and the Algonquin Radio Observatory in Ontario, Canada, as well as in the United States have tuned in fruitlessly to a tiny fraction of the stars in our Milky Way galaxy. Success would have been miraculous. Even if there are a million space messages coming our way, radio receivers would have to survey 100,000 stars to enjoy a fair statistical chance of detecting a single one.

All searches so far have been primarily for deliberate communications from space. Conceivably, radio telescopes might pick up unintended radio leakage from alien planets. Our routine radio and television transmission highlight Earth at broadcast frequencies. The antics of Lucille Ball have travelled more than twenty-five light-years out to the stars. Such unintentional terrestrial radiation increases every year and zips into space at light speed. The same phenomenon could disclose the existence of other technological societies to human eavesdroppers. Truly comprehensive searches should scan the sky for all possible radio signals.

SETI Proposals

A serious search for extraterrestrial intelligence (SETI) is feasible today with existing facilities! So reported Philip Morrison and NASA scientists John Billingham and John Wolfe in 1977, after two years of deliberations by distinguished scientists at a series of Science Workshops on Interstellar Communication conducted by the NASA Ames Research Center. Project SETI strategy is straightforward. It would attempt to tune into radio broadcasts from galactic civilizations for five years. It would conduct an all-sky survey plus a high-sensitivity targeted search of stars like our sun, using existing radio telescope and computer facilities.

For the all-sky search, the Deep Space Network antennas at Goldstone, California, would have multichannel analyzers to tune in four million different channels between 1400 and 2500 megahertz simultaneously! The giant Arecibo telescope is capable of detecting signals transmitted by its galactic counterpart at a target located thousands of light-years away, or less powerful local leakage similar to our television and radar. The Very Large Array is ready to receive even more distant or weaker artificial signals whether they are intentionally broadcast or not.

In the future if a more comprehensive search appears warranted, new construction could be undertaken. Project Cyclops is a detailed design for a superbly sensitive array of 1500 antennas each 330 feet (100 m) in diameter coupled to data processing systems to synthesize a collecting area of twenty-six square miles (65 square km). Another ambitious projected SETI design involves lofting very large, very lightweight antennas into space above interfering terrestrial transmissions. Future technology will surely evolve to follow the challenges opened by an inaugural Project SETI.

Efforts to make contact include "Sounds of Earth" recording which was sent into space mounted on Voyagers. Here, in Safe-1 Building at Kennedy Space Center, gold-plated copper disk (partially covered by flag) is being encapsulated in gilded aluminum shield to protect it from rigors of time and space erosion. Voyager is in background. Pictorial plaque was also used on Pioneer 10 and 11 to show location of Sun and planetary system, flight path of spacecraft and its size in relation to human forms, other fundamental data.

Puzzling Phenomena

Frequent reports of UFOs (Unidentified Flying Objects) from different parts of the world awaken speculation that extraterrestrials are visiting Earth. Some of these sightings are subsequently explained in familiar terms by authorities and so become IFOs (Identified Flying Objects). Other UFO reports continue to baffle even the top experts. There is no scientific proof that true UFOs are actually visitors from outer space. But there is no proof that they are not, either.

Imposing ancient ruins excite conjectures about past visits to Earth by superintelligent deep-space astronauts. Megaliths loom enigmatically at Stonehenge in England and Carnac in France. Monuments tower impressively in Egypt and Central America. It is not impossible that wise, capable creatures who were excited by gigantic construction projects descended from the stars eons ago. But archeoastronomers who investigate early astronomies through relics connect these silent stone records to the sky with much more thrilling theories.

UFOs

A true UFO is one reported by a normal individual who has had a personal perception of a completely unfamiliar light or object that remains unexplained after authoritative investigation. There are UFO reporters of both sexes, from all age groups and socioeconomic levels. Undeniably, anyone who reports sighting a UFO has had a real experience. But witnesses, who frequently are respected members of their community, are typically reluctant to face public ridicule by disclosing it. While some

UFO reporters may well be cranks or unstable individuals, there are too many others around the globe to simply dismiss.

Astronomer J. Allen Hynek organized the Center for UFO Studies in Evanston, Illinois, in 1973 to scrutinize UFO reports scientifically. At the Center he heads an organization of competent scientists who investigate and analyze thousands of submitted reports. Hynek classifies UFO sightings, which come in from around the world, into four basic categories: nocturnal lights, daylight disks, radar, and Close Encounters.

By far the largest number of UFO sightings occur in the dark of night. Thousands falling into the nocturnal lights category are reported annually. Typically they include a startling colored light of indeterminate size which moves—apparently guided—along an unnatural path. A bright light is the dominant feature and a supporting craft may be inferred but not actually seen.

Recorded nocturnal light observers include U.S. Air Force personnel, policemen, housewives, and students. Many of the lights they saw were fairly easily explained. No one knows just how many more would survive careful investigation by a team of unbiased experts. Since scientists are almost universally, militantly negative about UFOs being alien starships, they rarely investigate reports thoroughly. Still, some nocturnal lights defy natural explanation by interested specialists who understand the behavior of normal lights in the sky.

Disks or oval craft usually predominate in daylight sightings. Some turn out to be natural phe-

*Preceding pages: Two Close Encounters of
the First Kind. Both show domed Unidentified
Flying Objects photographed—sincerely
if not clearly—by witnesses with cameras
at hand. Close encounters are UFO
events reported by an observer within 180 m
of thing seen. First-kind is a sighting, nothing
more. True UFO is one reported by a normal
individual who has had personal perception
of an unfamiliar object that remains
unexplained after authoritative investigation.*

nomena or hoaxes, but hundreds of others remain unexplained. Observers frequently declare the colors and sounds they experienced are so totally foreign to daily life that both defy description. Usually, shiny metallic or glowing craft appear controlled, may hover a while, or speed away noiselessly.

Radar and corresponding visual sightings are reported by technically trained, reliable radar operators, pilots, and air traffic controllers. Radar and visual reports are not always compatible. Typically, a radar operator sees on the screen at night a strange blip unlike those due to large aircraft, weather effects, or ordinary equipment malfunctions. Correspondingly, a light or lights, with perhaps the suggestion of a craft, may be observed visually.

Close Encounters designate those that occur within about 200 yards (180 m) of the witness. Three varieties are most commonly reported. Close Encounters of the First Kind are simply sightings. Close Encounters of the Second Kind involve observable physical effects. Close Encounters of the Third Kind entail intelligent extraterrestrials.

In the most puzzling reports of Close Encounters of the First Kind, multiple witnesses testify they saw a bright or glowing light, and perhaps a small oval craft, hovering or accelerating abnormally. The craft may be capped with a dome. It never has conventional wings or wheels. Light and craft disappear without leaving any measurable physical effects behind. After obvious nonsensical reports are discarded, those remaining from multiple witnesses who are sincere, trustworthy people can hardly be dismissed as misperceptions or hallucinations.

Even more intriguing are Close Encounters of the Second Kind because they involve tangible effects on the environment or on living creatures which should be measurable. Marks supposedly left by alien craft may stay visible on the ground for days. Trees or plants may be destroyed. Interference with mechanical or electrical systems may occur and cars are sometimes reported as temporarily disabled. Bizarre animal behavior or temporary human paralysis or discomfort are other posited signs.

Amazed witnesses may report bright lights with perhaps a strange oval craft that flies in a very unnatural way. Here witnesses are usually not highly educated. Possibly sophisticated individuals refrain from making such reports for fear of almost certain ridicule by their equals. Scientists who want to conduct serious investigations of Close Encounter reports also face ridicule by their peers, together with an absence of sponsors.

Close Encounters of the Third Kind are the most extraordinary. Usually they involve humanoids who exit their craft and immediately adjust without difficulty to the surroundings. They have been reported by largely unsophisticated but evidently normal witnesses around the world. Some responsible citizens may describe beings in the alien craft but they rarely claim to have communicated. Reports of personal interactions between humanoids and witnesses usually turn out to be completely unreliable,

Left: Close Encounter of the First Kind sighted in 1965 in Exeter, New Hampshire. It left no physical trace of its dusk visit. It is unusual for a Close Encounter witness to have presence of mind to be able to photograph the UFO event. In many First Kind cases, multiple and reliable witnesses assert they saw glowing light and small, oval craft either hovering or accelerating abnormally. Craft may have a dome; never has conventional wings or wheels.

and all UFO phenomena are automatically derided.

Many UFO photographs exist, but none is absolutely authenticated. A number of purported UFO snapshots have been exposed as deliberate fakes, but others remain puzzling. Substantiation of an alleged UFO picture would require witnesses to its shooting. People do not ordinarily have cameras ready or the presence of mind to take pictures from various angles when they unexpectedly confront a frightening situation.

Most astronomers today consider extraterrestrials the least likely, although not an absolutely impossible, explanation for the thousands of recorded UFO reports. In 1977 the American Astronomical Society queried members who had spent as many as 365 hours studying UFOs. Only a few thought UFO investigation merited time and money that might be allocated to other research.

So far there are only UFO reports to investigate. There are no actual craft or parts to examine. There are no authenticated photographs of or interviews with extraterrestrials. UFO reporters are the only sensors of UFOs. The Center for UFO Studies must depend on their credibility in the absence of hard data. But the fact of multiple witnesses increases the credibility of a report, as does the high personal repute and normal mental health of the reporter.

Astronomer J. Allen Hynek has studied many interesting UFO reports and purported images. Obvious foolishness, natural occurrences, and otherwise explicable cases are weeded out. Still, reports from over 100 countries remain that have no univer-

UFO must leave physical traces indicating that it was at the precise site reported by the witness to be Close Encounter of the Second Kind. Top: Imprint of charred grass of unsighted UFO found in Odessa, Saskatchewan in 1977. Middle: 131 m/D shriveled soy bean patch in Van Horn, Iowa, on July 12, 1969, looked as though it had been subject to some sort of radiation. Bottom: Imprint found on day of UFO landing in Socorro, New Mexico, 4/24/64. Opposite: Matted grass at site in Langenberg, Saskatchewan where four vehicles, two side-by-side, were seen on September 1, 1974.

sally accepted interpretation. Hynek thinks there is enough data to warrant a systematic, rigorous study.

It is hard to conceive that every single UFO reporter is either misinterpreting the real world or hallucinating. Possibly UFOs may be something beyond the comprehension of modern science. The past offers too many examples of scientific smugness for practitioners to claim they have unlocked more than a fraction of the secrets of the universe. The UFO phenomena may require breakthroughs in our understanding of the cosmos that are beyond the wildest dreams of contemporary science.

Both UFO devotees and astronomers agree that UFO reports exist. Whether UFO sightings are something genuinely new to science or delusions must remain controversial until we have indisputable evidence.

IFOs

Many modern city dwellers are not conversant with the sky. The inexperienced may glance up casually, see a flashing light apparently approach-

ing, and mistake it for an alien craft if so inclined. Astronomers and space scientists often turn UFOs into IFOs by clarifying natural astronomical occurrences.

Refraction of the light from Venus as it passes through Earth's atmosphere can produce multicolored flashes and give the planet an oval appearance. Starlight is distorted by turbulent air and may shimmer in a weird manner. Small bits of matter from outer space burn up on shooting through Earth's atmosphere and leave light streaks called meteors. The largest, called fireballs, glow colorfully as their constituent chemical elements burn. They may move noiselessly, or noisily enough to irritate animals and make them visibly restless.

There are strange natural atmospheric effects that are documented though not well understood. Ball lightning has the form of a transitory, reddish, glowing ball up to about one foot (.3 m) in diameter. It may float or move rapidly along the ground before disintegrating. Swamp gas may produce spooky light as it rises from decaying vege-

tation. Swamp gas started a UFO bungle in Michigan in 1966 that prodded people to request a Congressional investigation of UFO phenomena in the state.

Meteorologists and military personnel solve other UFO mysteries. Their research balloons are lofted up to thirty miles (48 km) above ground for atmospheric measurements. The balloons can drift gently or shoot across the sky reflecting sunlight. Some shine lights of their own to avert collisions with aircraft. Glimpses of such balloons have triggered thousands of UFO reports.

Temperature inversions can send contorted hot air currents floating upward in disk-shaped clouds at sharply vertical angles. Light rays abnormally refracted by such layers of varyingly dense air create optical illusions that may call lighted craft to mind.

Flocks of birds or geese flying low in sparkling sunlight at dawn or dusk may resemble an alien squadron. Even swarms of insects can create bizarre effects. Entomologists Philip S. Callahan and R. W. Mankin of the U.S. Department of Agriculture carefully investigated eighty sightings of UFOs reported from 1965 to 1968 near Roosevelt, Utah. They subjected several different species of insects to laboratory-created electric fields comparable to those produced in thunderstorms, and the bugs lit up in brilliant colors. In 1978 the scientists announced that the Utah culprits were apparently a multitude of spruce budworms (a kind of moth) illuminated by a common electrical discharge through the atmosphere.

Some artificial satellites have metal foil coatings to reflect sunlight and to prevent overheating. They may radiate unexpected lights. The North

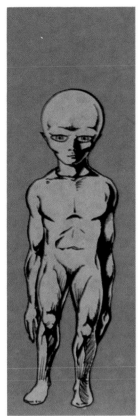 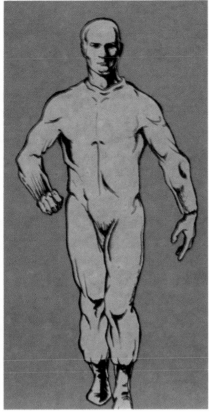

American Air Defense Command tracks thousands of objects besides working satellites in orbit around Earth. When one burns up on reentry into Earth's atmosphere, it may create a widespread scare.

Some UFO sightings are reported by people later unmasked as pranksters. Others are hoaxes deliberately promulgated by seekers of instant fame or fortune who try to capitalize on the widespread current interest in UFO sightings.

Psychologist Ernest R. Hilgard of Stanford University's Laboratory of Hypnosis Research in California suggests that UFOs, like hallucinations and near-death imagery, are mainly fabricated products of the imagination. If so, psychologists have yet to explain satisfactorily how multiple hallucinations can occur when there are confirming reports from several witnesses viewing the same phenomena from different locations.

Between 1947 and 1969 a U.S. Air Force UFO investigation called Project Blue Book identified 12,618 sightings. Another 701 were left unidentified, however. Project Blue Book was closed after twenty-two years of effort because the Air Force concluded that expenditure of further money and time was not warranted. UFOs did not appear dangerous and no purported alien spacecraft had even stayed long enough for laboratory analysis.

UFO devotees attempted to open another federal inquiry in 1977. In July, President Carter's science advisor, Dr. Frank Press, asked NASA to establish a small panel for inquiry into UFO reports. NASA Administrator Robert Frosch rejected the suggestion in December, reasoning that it would consume millions of dollars with little chance of significant findings. NASA did make a standing offer to analyze bonafide evidence any time it is preferred by credible sources.

A minority of scientists thinks UFOs are worth investigating in much greater depth than has ever been done. They point out that after throwing out nonsense and using all available explanations, some reports by multiple witnesses still cannot be explained. Even the most negative scientists would gladly examine a tangible piece of an extraterrestrial craft. Unfortunately, none has been found so far, and no report can be analyzed scientifically until at least one is.

A Close Encounter of the Third Kind involves a report of human contact with humanoid creatures. It is usually documented by drawings after the fact because witnesses are too distraught to photograph incident at moment it occurs, even if camera is available. Above: Creatures are similar to humans, although usually described as shorter, having an over-sized head, and with elongated eyes. Left: Rendering of appearance of humanoid creatures in Papua, New Guinea on June 26, 1926. Creatures were said to have waved back at natives from craft that never landed. Thirty witnesses were present.

Megalithic Ruins

Remarkable huge stone ruins in the British Isles, France, Spain, and Portugal are unexpectedly sophisticated. There are no known written records of the construction or purpose for these sites. Centuries of speculation have variously attributed them to Druids, Danes, Romans, Phoenicians, and even to ancient astronauts from other worlds. Modern astronomical interpretations of the standing stones and alignments reveal that prehistoric humans possessed unbelievable knowledge and capabilities!

Stonehenge on Salisbury Plain in southern England is probably the best known assemblage of prehistoric monoliths. Radiocarbon dating of artifacts there indicates that Stonehenge was built in several stages around 4,000 years ago. Archeological finds show that toward the end of the Stone Age humans were using polished stone and metal tools, kept livestock, and engaged in agriculture, fishing, pottery, weaving, and art in many communities. Various cultures contributed to Stonehenge for a thousand years.

Today at Stonehenge a broad approach called the Avenue leads in from the northeast. A 300-foot (90-m) circular ditch with a bank just inside surrounds the ruins. Within, immense stones are arranged in four startling series that could hardly have occurred by chance.

Huge upright sandstone boulders, or sarsen stones, form an outer circle that is slightly more than 100 feet (30 m) in diameter. Of the thirty original uprights, sixteen still tower almost fourteen feet (4 m) above the ground. Amazingly, each one weighs some twenty-five tons (23 t). The uprights have flat, smoothed inner sides and rough, unfinished outer sides. A continuous line of dovetailed lintels, each weighing about seven tons (6 t), apparently connected them. The sarsen stones probably originated near the site since similar ones are still found there.

A second circle almost seventy-seven feet (23 m) in diameter is formed inside by nine upright and eleven overturned bluestones. Similar bluestones are common in Wales which could have been the source of those at Stonehenge. The bluestones are especially colorful, with outstanding white spots when wet. They must have had very special meaning to the ancients who were sufficiently motivated to transport them a great distance.

A third series had five towering sarsen trilithons, or stone archways, in the shape of a horseshoe. Constructed about 1900 B.C., these were the largest at the site, and two are still standing. The colossal Southwest Trilithon is in the middle of the curve opposite the horseshoe opening. It has twenty-two-foot (7-m) uprights capped by a slab fifteen feet (4.5 m) long. Its massive western upright weighs some fifty tons (45 t).

Evidence from twelve remaining stones and excavations indicates that there was an innermost oval group. Inside the oval is a sixteen-foot (5-m) long stone that is now in two parts. Early investigators dubbed it the Altar Stone from its postulated religious use.

Large, rough upright stones are also found in fields and woods around the seaside resort of

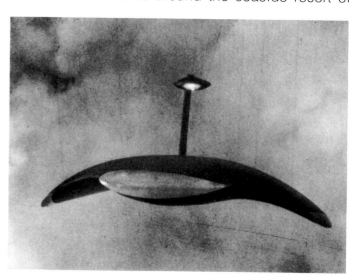

Carnac in Brittany in northwest France. Some are twenty feet (6 m) tall and weigh over a hundred tons. At Carnac there is a striking array of stones lined up eleven and twelve abreast that spreads over two and a half miles (4 km). Scattered stones to the west suggest that the original prehistoric complex there extended more than six miles (9.5 km).

The marvelous Grand Menhir Brisé near Carnac is the greatest building stone ever. When standing, it probably towered sixty feet in the air and weighed over 340 tons (300 t) ! Today the toppled stone is broken into five pieces, one of which is missing.

Stone monuments were apparently built during at least two millennia at different sites in Europe. Radiocarbon dating of relics such as bones, tools, arrowheads, cups, and beads reveals that the oldest monuments in Brittany were built 6,000 years ago.

Little is known about the farmers and fishermen who must have spent many backbreaking years transporting the megaliths to sites like Stonehenge and Carnac. Quite likely they were intensely interested in the cyclical motions of the sun and moon. These could form the basis of calendars important for agricultural tasks and tide forecasts. Eclipses of the sun and moon had special significance.

Conceivably Stone Age astronomers who could predict solar and lunar motions and eclipses accurately might have acquired enormous local political power. They might even have elicited from their people the requisite millions of strenuous work days for the construction of stone observatories.

Astronomer Gerald Hawkins aroused re-newed interest in the astronomical meaning of the Stonehenge alignments in 1963. Using computer analysis to back up his controversial theory, he pointed out that the precise construction and alignment of the stones facilitated the observation and prediction of significant motions of the sun and moon. An on-site survey by C. A. Newham, a British amateur astronomer and archaeologist, yielded a set of meaningful alignments with notable positions of the sun and moon. Apparently the creators of Stonehenge carefully kept track of sun and moon rises, sets and passages for many years.

Astronomer Fred Hoyle gave astronomical significance to Stonehenge in an interesting scheme of his own. The monument was again related to solar and lunar observations and eclipse predictions. Prehistoric builders were credited with sufficient knowledge of the sky to lay out a primitive observatory. Perhaps they worshipped powerful sun and moon gods. If ancient priests did use their astronomical knowledge for political gain, they might have ultimately lost power by overlooking the imperceptible changes in celestial motions that inevitably require continual correction.

Renowned archaeoastronomer Alexander Thom, a retired professor of engineering at Oxford University, surveyed the Stonehenge surface features in 1973, incorporating the best data from excavations. His careful measurements and interpretations of prehistoric stone rings and standing stones at over 300 sites have disclosed a number of lunar observatories. Thom convincingly demonstrates that aspects of Stonehenge evidence an as-

Film version of **War of the Worlds** *was notable principally for vivid special effects. Spaceships of extraterrestrial invaders looked vaguely like manta rays and caused indescribable havoc with atomic disintegration rays from goose-necked guns emerging topside. If aliens have reached Earth—or are about to—it could only be through solution of profound problems of time travel so far insuperable for us: a system of propulsion permitting speed near that of light, or life prolongation for travelers through cryogenics or induced hibernation.*

tonishing knowledge of basic geometry and astronomy on the part of its prehistoric builders.

All the monuments Thom analyzed were built using a common unit of measure which he dubbed a megalithic yard, equal to 2.72 feet (0.82 m). Stone Age people are thought to have been illiterate. Possibly the common construction techniques apparent at widely scattered ruins in Europe were exchanged by trained masons or architects who traveled hundreds of miles through different lands.

Thom interprets Stonehenge and the Carnac Alignments as precision astronomical observatories. The stones point to significant stages of sky crossings by the sun and moon. They evidence the potential for accurate calendar formulation and eclipse prediction. The Grand Menhir Brisé was seemingly erected to facilitate sightings from eight other stations in the area. A line of sight from these other structures to the Grand Menhir leads to horizon points marking extreme swings of the moon. Stonehenge is interpreted as the reverse. Ostensibly Stonehenge served as a universal backsight for several other stations in the area.

The exact astronomical meaning of the megalithic ruins is not yet certain. But that their originators were far more sophisticated than previously assumed is quite clear. To all appearances they understood and tracked the movements of celestial objects in precision observatories they constructed without the aid of any deep-space astronauts!

Pyramids and Temples

Colossal Egyptian pyramids and great temples still wow sightseers. Moderns find it astounding that the ancients were capable of building such structures. Some who are overwhelmed credit deep-space astronauts with the monuments.

Most impressive of all is the Great Pyramid of Cheops at Giza in northern Egypt. One of the Seven Wonders of the World, it is the largest pyramid ever built. Cheops reigned over ancient Egypt circa 2900 B.C. and was the founder of the fourth dynasty. His Great Pyramid is over 480 feet (144 m) high and covers thirteen acres (5 hectares).

Pyramids actually evolved as royal tombs about the time of the fourth dynasty. Each ruler built

Reports of UFOs are sometimes reclassified as IFOs because of clarifying natural astronomical occurrences. Opposite: Overseas U.S. Remote Air Force Tracking Station, one of a network of tracking stations spanning entire globe. Left: Balloon flight to study Earth's atmosphere at U.S. National Center for Atmospheric research at Palestine, Texas. Balloon, filled with helium and in flight, may reflect light from its surface, appearing to ground witness as UFO. Such glimpses have triggered thousands of UFO reports, when event is actually IFO.

his own to protect his mummified body and his treasures from human desecration and pillage for eternity. Egyptian pyramids are typically massive solid stone piles on a square base. Their triangular sides slope to an apex opposite each of the four principal points of the compass in an obviously significant design that is not yet decoded. An opening in the northern wall permits entrance. A small passage through lesser chambers leads to the burial chamber which is secured below ground level. It is certainly amazing that pyramids were erected without modern science and technology. Materials were assembled from distant sources as required. Copper came from mines in Sinai. Granite was transported from Aswan up the Nile.

Hieroglyphics show that the ancient Egyptians knew the sun, moon, and stars well. Astronomical imagery was basic to their mythology and religion. The sun and the star Sirius were specially noteworthy. Astronomers must have existed in even the earliest Egyptian communities, since they had the concept of a tropical calendar and twenty-four-hour day over 5,000 years ago.

Spectacular representations of the Egyptian sky showing planets and star groupings appear on tomb and temple ceilings. Mercury, Venus, Mars, Jupiter, and Saturn, the Orion constellation, and the Big Dipper are unmistakable. One of the most famous—the circular Dendera zodiac from the ceiling of the Temple of Hathor, circa 30 B.C., in Upper Egypt—is now in the Louvre in Paris. Hawkins has suggested that the alignments of the great temples are connected to motions of the sun and moon.

Astronomical interpretations of the pyramids and temples vary. The architecture incorporates such accurate alignments, which require precise astronomical measurements, they could hardly happen by chance. For example, the north and south sides of the Great Pyramid are aligned almost perfectly east-west. Its west side is practically north-south, and its east side just slightly west of north. Since the Great Pyramid was not an observatory, its astronomical orientation must have symbolized some very important but still unknown principle.

The Pyramid texts describe the dead Pharaoh s ascent to the heavens. Astronomer Edwin C.

Krupp conjectures that the ventilation shafts from the King's Chamber may have been oriented to conduct his spirit to the eternal cosmic cycle in the sky. They lead toward the never setting circumpolar stars in the northern sky or the Orion (Osiris) constellation and Sirius in the southern sky.

New World Wonders

Early Americans were also fascinated by the movements of the sun, moon, and stars. They built observatories and incorporated astronomical alignments into city architecture. Best of all, they left written records. The credit for their fantastic achievements cannot believably be assigned to deep-space astronauts!

Several cultural centers, with scientific and artistic accomplishments, flourished in Central America between 500 B.C. and 1500 A.D. The Mayas, members of tribes of Indians in Yucatan, British Honduras, and North Guatemala, and the Aztecs in Mexico had highly developed civilizations long before their conquest by Spain in the sixteenth century. Four original Mayan manuscripts, which luckily escaped burning by zealous Spanish conquerors, survive to reveal astounding astronomical sophistication. The Mayas had almanacs of the sun and moon, tables of the movement of Venus, and an accurate calendar based on astronomical cycles.

The pyramid-temples in Central Mexico might have been observation posts. Aztec records at the time of the sixteenth-century conquest refer to the practice of astronomical orientation of buildings. Astronomer Anthony F. Aveni has surveyed and interpreted architectural ruins at several sites in Central America. Still more remains to be explained, but he convincingly demonstrates direct relationships between underlying architectural alignments and sunrise or sunset positions on important agricultural, political, or religious dates.

Although they do not agree on the exact interpretation, several experts attribute the peculiar structure and orientation of the Caracol Tower at Chichen Itza on the Yucatan Peninsula to its function as an astronomical observatory. The Caracol is a circular tower that was constructed in units by the Mayas starting about 800 A.D. Entrance is gained through four outer doors which each face one of the principal compass directions. Inside, a circular corridor has four more doors, midway between these points. The top is reached by a spiral staircase.

At the top of the upper tower three windows remain which are obviously advantageous for astronomical observations above the flat Yucatan terrain. The Mayas utilized Venus, the sun, and the Pleiades star cluster for their precise calendar and the setting of dates for important events. Aveni surmises that the Caracol Tower was erected primarily to exhibit significant astronomical data through its architecture. Venus, the sun, and significant star groups could have been observed through its few windows.

Humans have long been avid stargazers, according to archeoastronomical investigations. Most likely the ancients studied the heavens both in wonder of the cosmos and in an effort to improve human existence. Space still beckons humans who look outward with the same dual motivation today.

Europe's ancient megaliths are believed to be ancient observatories of sun and moon, built during at least two millennia. Some people believe that England's Stonehenge (opposite) and France's Carnac (left) formed the basis of calendars, which were vital to Stone Age agriculture and tidal forecasts. Eclipses of sun and moon had special significance, so Stone Age astronomers able to predict solar and lunar motions and eclipses accurately may have wielded enormous power—enough to induce the seemingly superhuman efforts necessary to construct the megaliths.

Space Colonies

A permanent human community floating in space could be a reality early in the twenty-first century! Scientists have readied a design plan for a settlement off of the Earth, where people will breathe air, bask in sunshine, enjoy lakes, gardens, and woods without ever having to endure inclement weather, industrial pollution, or overpopulation. Humans have sufficient knowledge and resources to build and inhabit this space colony in only twenty-five years.

People have successfully lived on the moon and in space stations in orbit around Earth. Apparently there are no insurmountable hazards to prevent humans from living for long periods of time in space. Forseeable problems can be solved with available technology or through possible scientific advances. New challenges can be met as they occur. Space proffers adventure and treasure to the daring. Hopefully they will embark soon.

Rationale

The picture of Earth as a spherical space colony of finite resources became irresistibly persuasive as soon as the first space-age photographs of our planetary home were snapped. Knowledge of the burgeoning human population, limited supplies, and delicate ecological balance of planet Earth oppresses many people. In the past, opportunities for growth have fostered political freedoms for individuals, while scarcities have brought violent competition and tribulations.

Maintaining a vacuum and weightlessness on Earth is both expensive and requires sophisti-cated machinery. These and relative cleanliness in space cost nothing to produce and are available over vast distances. Thus, the potential applications of a space vacuum are impressive. They include producing components for better electronic systems, thin film silicon solar cells, and purer metals. With zero gravity, man will be able to produce microelectronic chips for computers and electronic devices. These will eventually be improved and made even more miniature. Also, new alloys from previously unmixable metals, larger and more perfect crystals, stronger magnets, new glasses for camera lenses, laser optics, improved microscopes, telescopes, pharmaceuticals, and welding procedures will all be possible.

There will be new industries with materials, processes and products unknown today, since space and its possibilities are still being explored. Thus, in the future the riches of space may be exploited in ways undreamed of today.

Conditions Out There

Pioneers must understand the physical properties of space if they are to successfully establish colonies there, for every aspect of their lives will be affected. Although space seems featureless, gravitation produces a topography as critical to prospective colonists as waterways, mountains, and valleys were to terrestrial settlers. A colony should be located at a site that gives the optimal balance between availability of resources, ease of access, rapidity of communication, and maintenance costs.

The availability of raw materials from the

Preceding pages: Feasible first colony wheels majestically in space. Some 10,000 people will live within tubular torus protected from lethal radiation by moonrock coating. Disk mirror (top) reflects light/energy of sun's rays. Above: Central sphere has docking tower for spaceships. Work sphere under it has rectangular heat radiator. Six spokes lead to habitat rim, contain elevators, power cables, pipes. Vanes between are solar panels.

planets, their moons, and the asteroids is determined by the energy required to lift the materials from their source. Every massive celestial body, such as a planet or moon, is effectively located at the bottom of a gravity valley. Nothing can get off the surface without expending energy. The more massive a world is, the deeper its gravity valley and the harder it is to escape.

Planets sit in deep gravity valleys. A space colony could not justify the effort required to transport materials from planets unless one is nearby or uniquely rich in indispensable supplies. Moons present less of a problem with their shallower gravity valleys. Our moon is encircled by a gravity valley only 1/22 as deep as Earth's. Logically, prospective colonists will plan to lift raw materials from the moon rather than Earth once they are established.

Earth's moon is an especially attractive source of metals such as aluminum, titanium, and iron for construction, oxygen for respiration and rocket fuel, and silicon for glass for an initial colony. The moon's resources, only modestly supplemented by necessities from Earth, could sustain human life and technology. Asteroids are even less massive than the moon and sit in shallower gravity valleys. In the future they could be tapped for hydrogen, carbon, nitrogen, minerals, and water ice.

At a point in space designated L_5, located on the moon's orbit equidistant from Earth and moon, the gravity and centrifugal forces of the two worlds balance each other. Here is an ideal gravity valley for the vanguard space colony. Once installed, it would stay in its practical midway location revolving, like Earth and moon, around the sun. Although the actual distance to L_5 is larger than to a geosynchronous orbit, the propulsive effort to get there is not significantly greater. Because of gravity, the main work is in getting off Earth's surface. Resources could be shipped in and out of L_5 with minimal expenditure of propulsion mass, colony maintenance costs could be kept low, and transportation time would be reasonably short.

Radiant energy permeates apparently empty space. Sunshine is one of the primary resources colonists can utilize above Earth's atmosphere. An average square meter of space receives almost eight times more solar energy than does a comparable area on Earth. Sunlight can be collected for use in the colony with plenty left over for conversion to electricity for the energy-starved home planet. The vacuum itself will be extremely useful in manufacturing processes that are best accomplished in the absence of air.

Dangers threatening human life in space differ from those confronted by terrestrial pioneers. The fierce intensity and penetrating wavelengths of sunshine unattenuated by an atmosphere are serious perils. In addition, cosmic rays from the galaxy are a continuous source of lethal radiation.

Extraordinarily large solar flares occur every few decades. A solar flare is an activity where high-energy protons traveling near the speed of light are spewed into space. The high-energy protons could deliver many times the radiation dose known to be fatal to unprotected humans in less than an hour. Current optical and radio monitors give only a few

Assembly Non-Tethered Ship, operating somewhat like fork lift, conveys louvers to colony for installation by free-floating construction crew. Louvers absorb cosmic radiation while allowing sunlight to be reflected inside. They may have been manufactured on Earth, on a moon where metals are extracted and fabricated, or in factory of neighboring colony. Shuttle (background) is workhorse delivery truck in space. Mirror at top is for solar-power generation. This colony may be in operation soon.

minutes warning of an outburst before the arrival of the peak of the proton flux. The energetic particles can persist for more than a day in all directions.

Meteoroids are common, but are expected to be relatively harmless. The composition, distribution, and frequency of meteoroids have been measured on the moon and in space. A typical square in space one kilometer across is hit by a one-gram meteoroid about once every ten years and by a 100-gram meteoroid about once every 5,000 years on the average.

The probability of a large meteoroid crashing into a space colony seems comfortably negligible. Further, there is little likelihood that a small meteoroid will cause severe structural damage on impact. Still, the blast effects of even a tiny meteoroid could be serious, if it lost all its energy in a direct strike. Hence, colonists must plan to shield their habitat against meteoroid collisions as well as radiation. Small meteoroids could conceivably puncture holes in the habitat, causing slow leakage of air. A regular safety inspection and repair program within the colony would avoid any fatal leaks.

Humans today have the capability and resources to safely live in space and to exploit it for the benefit of all humanity. Cultural evolution on Earth has been a process of freeing ourselves from the constraints of the natural physical environment. Artificial environments commonly add comfort to our daily lives. They provide protection for people in spacecraft orbiting our planet. A hollow world like Earth with air, sunshine, water, and vegetation suspended in space is almost certainly the next step.

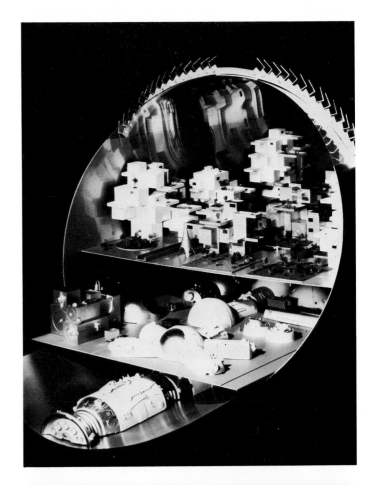

Top: Under living area are two service levels providing power distribution, storage, and perhaps processing equipment for light industry. Atop cylinder are chevron shields to reflect sunlight through baffles while deflecting harmful cosmic rays. Shell of torus is aluminum. Bottom: Space factory is sited several miles away to minimize effect of accidents and avoid polluting habitat. Major activities are construction of new colonies and solar-power stations, plus repair of robot satellites in Earth orbit.

The First Colony

A 1977 NASA report "Space Settlements, A Design Study" (NASA SP-413) details a feasible first colony that could be built and comfortably inhabited in space within twenty-five years after it is funded. Its creators were thirty engineers, physical and social scientists, and architects who had met and analyzed the physical, psychological, and social requirements of a permanent human community in space. Physicist Gerard K. O'Neill of Princeton University, noted for innovative ideas and calculations on space colonization, was technical director.

The envisioned space colony provides a permanent earthlike community in which 10,000 people work, raise families, and live fruitful lives. It is sufficiently productive to obtain and process enough raw materials to maintain itself. It utilizes everything space offers fully, and even produces goods and services for trade with the home planet Earth.

This space colony looks like a giant wheel. It is formed by bending a 427-foot- (130-m-) wide hollow tube into a huge torus (doughnut shape) that is over a mile (1800 m) in diameter. People live, work, and play in the torus. At its center is a hub where incoming spacecraft dock. Six large spokes each forty-eight feet (15 m) in diameter connect the hub and wheel. They have elevators for entry and exit to the living and agricultural areas in the torus, power cables, and heat-exchange pipes.

Glass windows mounted on aluminum ribs cover a third of the torus to let in sunlight. The rest of the torus shell is made of aluminum plates. A separate rough outer shell around the torus made of rubble from the moon protects the colonists from cosmic radiation. Shields over the windows pass light but block dangerous radiation.

The whole habitat rotates inside the outer shield at the rate of one revolution per minute around the central hub. The steady spinning provides comfortable artificial gravity in the torus. As the outer edge turns, a strong centrifugal force is created there. This is a happy substitute for Earth's normal gravity, which is currently assumed to be essential for physical and mental health.

Natural sunshine illuminates much of the colony's interior and supplies all of its energy. A big stationary mirror is suspended directly over the hub. Inclined at a 45-degree angle to the axis, this mirror reflects sunlight onto other mirrors which then direct it into the interior through a set of louvered mirrors that pass light but not cosmic radiation. Silicon solar cells are suspended over the spokes to supply fifty megawatts of electric power.

A 330-foot- (100-m-) diameter sphere on the side of the hub is a fabrication facility. Here metals are shaped and formed and a great deal of assembly and construction work is done. A 200-megawatt solar power plant and industrial furnace are on one side of the fabrication sphere. A large paddlelike structure is set below the hub in the opposite direction to radiate waste heat into space.

The space colony is ideally located in space at L_5. Here there is abundant sunlight, a weightless environment, and reasonable access to lunar minerals and Earth's capital and markets.

Construction

Designers think that the United States, possibly in cooperation with other nations, should initiate construction on the NASA prototype space colony or a similar one now. To make the design a reality, critical gaps in present knowledge must be filled through extensive basic research. Physiological and psychological limitations of an isolated population and the dynamics of closed ecological systems are being investigated. Testing and evaluation take place on Earth, in spacecraft in Earth orbit, and later on the moon.

In the designers' scenario, the first essential components of the colony are constructed from terrestrial materials by crews from Earth. Later, materials are mined on the moon and transported to the colony for processing and construction. A space station in orbit around the Earth serves as crew quarters, construction shack, and supply depot while astronauts assemble modules for the colony and for a lunar base.

Special attention is devoted to quickly establishing an early base on the moon. It has a nuclear power station, transportation system, crew quarters, and a processing plant. The lunar crew of 150 people requires 280 tons (250 t) of supplies from Earth and the rotation of seventy-five people annually.

At the outset colonists, supplies, and raw materials are transported to the colony from Earth. An atmosphere, water, chemical systems, and the first agricultural biomass complete the initial outfitting while the colony is in the build-up phase. Re-

Living in Space: Cutaway drawing shows organization of Earthlike community inside torus. Great wheel is 1,800 m/D, spins on axis once a minute to create artificial gravity for colonists. Cross-section seen here is 130 m/D. Sunshine illuminates tubular city. Environment is controlled to permit days, nights, seasons. There are no skyscrapers, freeways, or cars. Terraced, modular housing makes maximum use of area. Gardens and trees soften view. Moon can provide building materials. Note spaceship between spoke and rim.

placement supplies from Earth are then handled at the rate of 1,100 tons (1000 t) a year. An average of fifty new people together with personal belongings and the additional carbon, nitrogen, and hydrogen needed to sustain them in space are transported weekly to the colony. Oxygen and other elements are hauled from the moon. After regular immigration of colonists starts and the lunar base is operational, only those things not available in the colony or from the moon need be shuttled from Earth.

NASA's projected mission timetable schedules five years of research on Earth, three to five years for development and testing in orbit around Earth, five years to build up operations on the moon and at L_5, six years for habitat construction, and a final four years for completion of the shield and immigration of colonists. The total cost of the colony is figured at $190.8 billion in 1975 dollars.

Frontier Life

In the space colony the best of Earth's environment is reproduced artificially and the worst avoided. It is a tiny paradise in the dark void. Inhabitants enjoy the thrills and satisfactions of opening a new frontier. As the community grows, there is an exciting sense of dynamic change and limitless possibilities for self-fulfillment.

At first the colony fills up at the rate of about 2,000 people a year. New arrivals land at one of the docking ports at the hub. Near the central axis the rotation rate of one revolution per minute does not produce appreciable artificial gravity, so they feel weightless. The newcomers enter an elevator in one of the spokes for the 920-yard (830-m) trip out to the torus. Thousands of commuters travel in the elevators daily to and from their work in the fabrication sphere or outside the habitat. As the elevators move outward a sense of gravity begins, so passengers feel they are descending. In this world, outward is down.

The elevator takes people to a busy city. Bright sunlight streams on it from far overhead, illuminating a delightful scene. Homes, walkways beautifully decorated with bright flowers, frequent clusters of small fruit trees, parks, and bicycles are here. There are no skyscrapers, freeways, or traffic.

There are only 106 acres (43 hectares) available to house 10,000 people. The torus is divided into multiple layers to optimize the utilization of available space. A feeling of spaciousness is achieved by terracing structures up the curved walls of the torus and by locating shops, light industry, and mechanical subsystems below the central inhabited plain where artificial illumination is provided. Homes have large window areas to provide a sense of openness.

Housing is modular, permitting structures as high as four and five stories. Weather is never inclement in this controlled environment, so walls and doors are necessary merely for privacy. Each colonist is allocated 470 square feet (43 m²) for residential and community life.

Homes are compact but completely furnished to be convenient and attractive. Many apartments have balconies with flourishing plants. Furniture and the few ornaments are made of aluminum

All food is produced in three multitiered agricultural zones with total area of only 110 acres (44 ha). Separation from living spaces permits higher temperatures, humidity, CO_2 levels, and illumination for rapid growth. Fruits, vegetables, and grain thrive in well-ordered plots irrigated by river from park area at rear and indirectly by fish tanks at top level on either side. Rabbits and 2,800 cattle also would be raised here. Bottom level is a drying facility. Yield would be far greater than for comparable area on Earth.

and ceramics. There are hardly any woods or plastics, because these would have to be imported from Earth. In addition to home space each colonist is alloted forty-five square feet (4 m²) for mechanical and life-support subsystems and 220 square feet (20 m²) for agriculture and food processing.

Because of the frontier life and labor needs of the first colony, its first population consists mainly of able-bodied men and women with appropriate skills. Some children come with their parents and others are born in the colony. Construction workers with their immediate families are the first settlers.

The population ages and becomes more like that of Earth with the passage of time. People have a sense of mission and purpose that makes the colony a truly rewarding place to live. There are no marked socioeconomic differentiations but rather a frontier spirit of equality; every job is an important one. Since each individual colonist is selected as a settler only after rigorous examination, the whole colony shares a proud sense of elitism.

Food in the colony is nutritious, ample, and appealing. There is sufficient water to sustain life and to maintain a high level of sanitation. The colony's agriculture supplies the average person with about 2,500 calories and two quarts (2 l) of water in food and drinks daily. Waste generation and reduction are carefully balanced, so recycling eliminates the need to replace materials with expensive imports from Earth.

Three controlled agricultural zones are separated from three residential zones within the torus. Farmers utilize this arrangement to employ higher

than normal temperatures, carbon-dioxide levels, humidity and illumination to force rapid growth. Separation also inhibits the spread of disease. The system feeds the entire colony on the produce of only 110 acres (44 ha).

Productive farms raise fruits such as apples for pie and vegetables like carrots. Foods have acknowledged psychological importance and are selected for their cultural as well as nutritional importance. Each culture has favorites that are transplanted with settlers. Diets will no doubt be more international when the colonies are. This settlement was designed by Americans. Ponds hold thousands of fish. Their water is also used to irrigate fields of corn, sorghum, soy beans, rice, alfalfa, and vegetables. Crops such as wheat and tomatoes flourish on multiple tiers. Harvests are sequential to ensure continuous supplies. Meat comes from cat-

Opposite: Pair of cylindrical colonies 32 km long and 6,400 m/D, as seen from spaceship 32 km distant. Depending on interior plan, each could house 200,000 to several million people. At top are manufacturing and power facilities. Cups in ring are agricultural stations. Fold-out mirror panels regulate seasons. Left: Inside cylinder. Rotation on axis every 114 seconds creates gravity on perimeter (thus holding water in foreground pool). Above: Bridge the size of San Francisco's Golden Gate fits comfortably within colony landscape.

tle, chickens, and rabbits bred in the colony.

The colony's atmosphere can safely and comfortably support human life at a lower pressure than is normal at sea level. It contains enough oxygen to maintain good respiration. Nitrogen is added for its many benefits, including prevention of unusual body decompression, as well as to provide a safety margin during accidental pressure drops. Carbon dioxide is present for photosynthesis although it is kept below toxic levels. A comfortable relative humidity and temperature are maintained.

In the colony's controlled environment, sunlight, carbon dioxide, and chemical nutrients are harnessed to produce vegetation and from that to raise animals. The oxygen and water vapor which are released as by-products regenerate the atmosphere. All waste products from plants, animals, and humans are recycled. Waste processing restores carbon dioxide to the atmosphere, reclaims plant and animal nutrients, and extracts water.

Water is essential for drinking, humidity control, irrigation, waste management, and fire protection. The colony's water supply system is engineered so that more than enough is always available. Recirculating showers, low volume lavatories and efficient water usage in food preparation and waste disposal are employed.

The colony abounds with cultural and recreational activities. There are a fine library, schools, theatres, movies, and sports areas. Favorite ball games are quite different from their usual forms on Earth. The low gravity and rotation makes balls fly longer, faster, and along curved paths. Old games

are adapted to the new setting by instituting new rules and tactics. Sunbathing is unchanged and even a popular form of relaxation, as is television that brings news and entertainment into homes.

The first colony will never have to become overcrowded. If it is as successful as expected, the colonists will construct new ones. Our solar system has ample sunshine for energy and great quantities of untapped matter. More and more colonies can be built until our entire solar system is populated.

Commerce

Colonists are concerned with increasing their standard of living, becoming self-sufficient, and increasing their mastery of the space environment. Their industry attracts investment capital from Earth. A favorable balance of trade is maintained with the home planet. Construction is the major export, while products and services of highly developed technology as well as carbon, nitrogen, and oxygen are imported.

Because the community is small and isolated, there is great interest in establishing new colonies. Construction of new settlements is the habitat's major industry. Its second industry is the manufacture of satellite solar-power stations that capture solar energy while orbiting Earth and transmit it to Earth for conversion to badly needed electric power. Another lucrative activity is the repair and maintenance of robot satellites in Earth orbit.

Manufacturing and scientific and medical research proceed actively in the high-vacuum, low-gravity hub area. The fabrication sphere is placed

outside the habitat south of the hub to avoid industrial pollution of the atmosphere as well as to isolate a possible industrial accident. The plant is operated remotely to maximally exploit the vacuum of space. It has its own solar furnaces and electric power station run by solar energy. Maintenance is routinely performed in a normal environment in small spheres attached to the plant.

As the construction of multiple colonies ensues, rapid transportation between habitats is developed. The settlements support themselves by manufacturing space structures and supplying power to Earth. Mining operations expand from the moon to the asteroids and beyond. After some years, the many colonial communities are as populated and prosperous as the home planet.

To the Stars

The first great step into space has already been taken. Human footprints are on the moon. Inevitably space colonization will occur shortly unless Earth is destroyed through human folly. If the first colony is successful, ingenious pioneers will build other worlds farther and farther from home. Humans could survive in artificial environments throughout our solar system. Far in the future, when interstellar travel and settlements on other worlds are common, they may colonize other solar systems.

Modern science and technology can practically create a new world to order in space. Now the obstacles to human survival and expansion toward the stars are principally philosophical and social. Let us move outward courageously!

Two views of another self-contained space colony. Central sphere is 1.6 km in circumference. On inner surface, population of 10,000 lives and works, as in cutaway drawing at bottom. Sphere is circled by solar panels. Docking and zero-gravity industrial areas with independent solar-power plants are at either end. Blanket of slag from processing of lunar materials aboard colony shields sphere from cosmic bombardment. Whatever its vehicles, humanity can now embark on tremendous adventure of colonizing the universe.

Index

Italic numbers refer to illustrations

Picture Credits